THE GENE
CIVILIZATION

FRANÇOIS

GROS

THE GENE
CIVILIZATION

McGraw-Hill, Inc.

New York St. Louis San Francisco Auckland Bogotá
Caracas Hamburg Lisbon London Madrid
Mexico Milan Montreal New Delhi Paris
San Juan São Paulo Singapore
Sydney Tokyo Toronto

English Language Edition

Translated by Lee F. Scanlon
in collaboration with
The Language Service, Inc.
Poughkeepsie, New York

Typography by AB Typesetting
Poughkeepsie, New York

Library of Congress Cataloging-in-Publication Data
Gros, François, 1925 –.
  [La Civilisation du gène. English]
  The gene civilization/François Gros.
      p. cm. — (The McGraw-Hill horizons of science series)
  Translation of: La Civilisation du gène.
  Includes bibliographical references.
  ISBN 0-07-024963-6
  1. Molecular genetics. 2. Human molecular genetics.
3. Biotechnology, I. Title. II. Series.
QH442.G7913   1992
574.87 328—dc20   91-29174

The original French language edition of this book
was published as *La Civilisation du gène*, copyright © 1989,
Hachette, Paris, France.
Series editor: Dominique Lecourt

TABLE OF CONTENTS

INTRODUCTION

We have arrived. In less than three decades, molecular biology has, step by step, overturned the entire structure of the life sciences—from the theory of heredity, of course, through plant physiology to embryology and the neurosciences. In a few short years, this upheaval in the order of knowledge has come to be reflected in pathology and medicine, from cancer research to the study of human reproduction and aging.

And now, before our eyes, not only public health but agriculture, animal husbandry and whole sectors of major industries see unexpected prospects opening up before them. The marketing of bioproducts will undoubtedly alter, in one direction or another, the balance of the world market, and particularly the relationships between the industrialized countries and the South. The day is also at hand when we will have achieved technical mastery of our own bloodlines, and soon of speciation.

Yes, we have arrived. A gene civilization is taking shape around us. There is a need, and one which brooks no delay, for us to grasp the import of this maor development so that we can derive the best from it, failing which we shall very soon have to suffer its worst consequences.

That is the essential message of this book. It offers the reader, in a few pages of utmost clarity, the history of molecular biology, a picture of the biotechnological revolution and a statement of the political, juridical and moral issues that go hand in hand with the emergence of this civilization.

The gripping history recounted for us here is one of lightning-fast progress. While its origins, both near and remote, lie in the speculations of philosophers and the studies of naturalists, from Aristotle to Lamarck, its real beginning came only with the formulation of Darwin's theory of natural selection, which laid to rest the concepts of the scale of being and the economy of nature.

By freeing himself from these theologically inherited associations which still dominated the thought of the great comparative naturalists of the 18th century, Charles Darwin discovered in the order of living things a hitherto unknown type of causality. The essentials of his theory are known: small variations emerge in individuals and, when these confer an advantage in the struggle for mastery of the environment, they are transmitted to their descendants; the result, through selection and accumulation, is progressive modification of species.

It is apparent that this concept of causality is not only mechanistic: it also excludes any purposiveness, despite the ambiguities of the term "selection." Faced with misunderstanding on the part of his opponents, and often on the part of his adherents as well, Darwin was compelled to return unceasingly to this point: selection is not a power of choice; nature is not working towards an objective it has set itself in advance. Given variability, selection is nothing more than the necessary effect of vital competition.

INTRODUCTION

That was the starting point. It can be dated from the publication of *The Origin of Species* in 1859. But once this start had been made, to extend Darwin and, for example, answer the questions he had left unanswered, in accordance with a linear scheme of the history of science which is still too often accepted, was not enough. On the contrary, the history of research into living matter, which as we know has led to spectacular successes, paradoxically consisted in part in a reaction *against* Darwin, even though its result, in our times, was to confirm the essentials of the theory of natural selection.

To understand this, it must be recalled that the theory of natural selection is one thing, and that the idea of evolution is another. This idea had long predated Darwin's work, and had only just recently been restated by the English philosopher Herbert Spencer. In 1851, extrapolating from the recent work of the Russian embryologist of Prussian origin, Karl Ernst von Baer, Spencer had proclaimed a general law of evolution according to which "all organic development [was] a transition from the homogeneous to the heterogeneous." This law was held to apply to all of the life sciences as well as to psychology, history, and even cosmology. A full-fledged ideology, the ideology of evolutionism, was thus systematically erected on this foundation.

It was no doubt in particular because it had helped overthrow the dogma of fixity of species that Darwin himself was not fully able to disabuse himself of this idea, or rather of this ideology, and ultimately allowed himself, on certain points, to be caught in the trap of a way of thinking which was in fact profoundly purposive, since it quite squarely attributed an objective to the march of nature.

Thus, when Darwin brought his thinking to bear on the sensitive area of human paleontology (*The Ascent of Man*, 1871, deplorably translated into French as *La descendance de l'homme* [The descent of man]), the problem of races and psychology, it was as an evolutionist that he spoke. Still more serious, this purposiveness underlay even the hypothesis he formed of the causes of variability and the mechanisms of hereditary transmission, a process of which he was unaware: this hypothesis that expressly excluded any leap, and hence any mutation, and endorsed the classic image of transmission as a merger or mixture of the two parent stocks.

Of Darwin's thought, as of that of many great thinkers, it could thus be said that it did not succeed in fully keeping pace with its own novelty.

The great German zoologist Ernst Haeckel, who from 1860 until his death in 1919 made himself the untiring propagandist of a Darwinian version of evolutionism in the service of a general theory of progress, set Darwinism more firmly on its wrong course, giving rise to a very far-reaching historical misconception. Suffice it to recall that his book on *The Enigmas of the Universe* [*Welträtsel*, published in 1899], sold four hundred thousand copies in German and was immediately translated into a host of other languages.

The body of Haeckel's monumental work is also structured around the proclamation of a law, known as the fundamental biogenetic law, which posits, drawing its general inspiration from Spencer, that "the series of forms through which the individual organism passes, from the primordial cell to its full development, is simply a repetition in miniature of the long series of transformations undergone by the ancestors of the same organism from the most remote times to the present." In short, in the words of an aphorism which has remained famous, "ontogeny...is a short and quick recapitulation of phylogeny." The development of the individual reproduces in condensed form that of the species.

This evolutionist ideology helped to blind the best naturalists when they had before their eyes, in 1865, or less than ten years after the publication of *The Origin of Species*, the results obtained by Gregor Mendel on the hybridization of peas, set forth in his *Experiments in Plant Hybridization*. The questions raised by Mendel regarding the distribution of distinct "characters" (smooth, wrinkled, green, yellow peas, and the like) from generation to generation were quite simply unthinkable within the framework of that ideology, and the mathematical treatment he gave of them was bound to appear incongruous.

It took half a century for the truth established by Mendel to be "rediscovered" by Hugo de Vries, William Bateson, Thomas H. Morgan and a few others. In other words, for fifty years this truth simply did not exist; it was unable to give rise to any of the effects of research and experimentation which surround any scientific truth and define it as such.

Must science therefore be said to have fallen behind schedule?

The pace of scientific progress is in fact so far from linear that it cannot be discussed in terms of earliness or lateness without the risk of falling into caricature.

The evolutionist ideology did indeed have a mystifying effect, misleading even the very greatest minds of this century. Witness Freud, who from *Three Essays on the Theory of Sexuality* (1905) to *Civilization and its Discontents* (1929) cites Haeckel's biogenetic law to back up some of the key concepts of his doctrine (sublimation and repression, among others). On the other hand, it must be added that, by anticipating the unification of the various branches of biology (paleontology, embryology, physiology and the theory of heredity), evolutionism opened the way, or lent impetus, to research work without which molecular biology would never have come into being, or at least would not have expanded its field of application as fast as it is doing today. It even inspired research, outside the strict sphere of the life sciences, into animal behavior (George Romanes) and the mental development of the child (Wilhelm Preyer, James Baldwin), which laid the foundations for contemporary ethology (Konrad Lorenz) and child psychology (Jean Piaget).

Nevertheless, the fact remains that there was no possible way in which the evolutionist ideology could be synthesized with Mendelian mutationism. How was the concept of mutation to be incorporated into the recapitulative scheme of the biogenetic law, which saw everything in terms of progress, that is, in the first place in terms of continuity?

By a new "trick" of history, it was finally out of Mendelism and its extension, passing through several reformulations, to population genetics and then to molecular biology, that confirmation of the Darwinian theory of natural selection emerged, but a confirmation which at the same time definitively freed it from the evolutionist matrix in which it had been embedded.

This was the beginning of another history, the one recounted in this book. François Gros, who played his part in it, gives an admirable description of the emergence and expansion of molecular biology. I will thus confine myself to pointing out that this history once again defies any linear scheme, and indeed any predetermined "logic."

For molecular biology to be possible, genetics not only had to revert to its Mendelian "stage," but two other lines of research, originally completely independent, also had to be developed: the cell theory (Rudolf Virchow, "every cell comes from another cell," 1849) had to gain acceptance and a genuine chemistry of living things had to be developed, notably through Louis Pasteur's work on fermentation from 1854 onwards. Above all, it was essential for the physicists, with Erwin Schrödinger in 1945, to establish an unexpected link between information theory, as it was then emerging, and the knowledge that existed of life at the cellular level, thus providing the gentle push that launched the movement.

But it must also be noted that although the field of molecular biology was opened up in this way, its borrowings from information theory very quickly— by the end of the 1960s—came to present an obstacle to research. Like natural selection before it, the genetic code was enlisted, so to speak, in the service of a preexisting and all-explanatory ideology, structuralism, which drew on analogies between the coded structure of the genetic material, the structure of language, structures of kinship, the structure of the unconscious and even of the modes of production, in an attempt, with comparable repercussions, to unify the whole range of scientific disciplines on a single theoretical basis.

Like evolutionism in its time, structuralist ideology, whatever may be said of it today, did not produce only mystifying effects. Nevertheless, a profound "crisis of confidence," as François Gros terms it, had to occur among the ranks of biologists in order to forge the conceptual tools with which a bold "ascent" from the study of bacteria to the study of higher organisms and the exploration of the major functions of living matter could be initiated and carried through.

That the successes achieved in basic research since that time have more and more rapidly come to have increasingly wide-ranging practical implications is apparent from the impressive picture of the biotechnological revolution that so graphically emerges from these pages. Former director of the Pasteur Institute and afterwards scientific adviser to the French Prime Ministers Pierre Mauroy and Laurent Fabius, François Gros, who has thus over the past few years played, in more than one capacity, an active part in this revolution, gives the reader all the information and prediction data available worldwide. Who could do it better?

Throughout these pages, a query takes shape which is explicitly stated in the last two chapters: what kind of answer can biologists give, as biologists, to the questions now being insistently asked of them by average citizens and decision-makers alike?

These questions are, we know, inspired as much by hope as by concern. François Gros fully illuminates the paradox and the extreme difficulty of the situation: research into the human genome will make it possible to know and combat the most terrible of hereditary diseases; but the same research can be placed in the service of a comprehensive system of labeling, and lead to what he calls a "systematic genetic ostracism"—of which examples can already be found—engaged in by employment agencies and immigration departments.

The same very precise techniques of prenatal diagnosis and *in vitro* fertilization which offer the prospect of preventing some of the most calamitous human disasters also abruptly revive, in a molecular version, the fearsome spectre of the eugenics movement of recent tragic memory.

Let me pause here for a moment to make this last point fully, since it was central to the first description given, with a view to inspiring fear, of what a "gene civilization" might be. It will be recalled that it is a "foolproof system of eugenics," which governs the standardized and hierarchized organized life of the "brave new world" imagined in 1932 by the writer Aldous Huxley, son of the great biologist Thomas Huxley and brother of Julian Huxley, one of the founders of the evolutionary synthesis.

In the same year in which his brother became the first Director-General of UNESCO and stated that "we must increasingly rely on raising the genetic level of man's intellectual and manual abilities" in order to resolve the problems of education, Aldous Huxley wrote in a new foreword to his famous novel that "the only really revolutionary revolution [would be] that involving the

application to human beings of future research in biology."
We are all aware of the sinister picture he gives of the antic-
ipated social effects of this "ultimate, personal revolution."

At the time, the terms "eugenism" and "eugenics" had
been in current usage for quite a number of years. They
were invented and first defined in 1883 by Francis Galton,
Darwin's cousin and the father of differential psychology
and biometry. Galton considered that natural selection
played a negative role in society, noting that the working
classes had more children that the leisure classes. He saw
in this a threat to the western nations in their competition
with the rest of the world. Hence the program for "developing the most gifted races." In the United States, in England,
in Germany, eugenics societies soon sprang up to study by
statistical means recommended by Galton himself, and then
by Karl Pearson, his disciple and spiritual heir, the heredity
of mental diseases, alcoholism, etc. These institutes sought
to encourage the birthrate among the higher classes of society through a system of elitist family allowances, and to
promote research into artificial insemination. They were
among the instigators in the United States of the 1924
Immigration Act, which set quotas for immigration from
Central and Mediterranean Europe. They also encouraged
campaigns for mass sterilization of the "abnormal," whose
victims in several American states a few years later were to
be numbered in the tens of thousands.

François Gros twice mentions the book by Dr. Müller-
Hill (*Tödliche Wissenschaft* [Deadly science], Rheinbeck,
1984) which shows how the spectre of eugenics was
exploited by the Nazis in the service of the "final solution"

and the improvement of the "Aryan race." But, though its tone is indeed different, the position taken by Julian Huxley immediately after this horror, like the worldwide interest aroused by the recent affair of the conservation of Nobel Prize winners' sperm, with William Shockley as its protagonist, show that the spectre transcends frontiers and time.

François Gros' call for vigilance is understandable, for the means by which we are able to control and master the genetic machinery are of a power and a precision that cannot be measured on the same scale as those available in the golden age, if so it may be termed, of eugenics. But what he shows also is that, in essence, the question is not one of biology; it is a question of politics, law and morality. In short, a question of responsibility.

And herein lies what is not of least interest in this slim volume: a biologist who is among the greatest and the most committed members of his profession and yet rejects biologism. While he establishes that certain serious social issues are attributable to biology's transformation into a "hands-on science," while he emphasizes that the further development of certain lines of research may help resolve these issues, he rejects the idea that for every social problem there is a biological answer.

Yes, I repeat: we have arrived. But the gene civilization is not the best of all worlds. We are not being swept along by any tide of fate. A movement has begun: it is for us to know how to draw from it the best for our world.

Dominique LECOURT

I

MOLECULAR

BIOLOGY

FROM DARWIN TO
THE BIRTH OF GENETICS

In order to appreciate the scope of the vast cultural, economic and social upheaval that has resulted from the recent and rapid advances in the life sciences, we need to know the history from which these advances have emerged. Essentially, this history begins with the emergence of molecular biology. Its broad features will be outlined here.

Even if its roots extend down into a distant past illuminated, in various respects, by the names of Aristotle and Galen, Vesalius and Descartes, and also those of the great comparative naturalists of the 17th century and the Age of Enlightenment (Linnaeus, Adanson, Buffon, Cuvier, Geoffroy Saint-Hilaire and so many others), the Darwinian theory of natural selection may be said to be its most decisive theoretical and historical prerequisite, and still the most active today.

The discontinuous, rigid and sacrosanct representation of species was of course severely shaken by the work of Jean-Baptiste Lamarck, whose speech on May 11, 1800, is a milestone in the history of biology, and whose *Zoological Philosophy*, published nine years later, remains one of its finest monuments. But with Charles Darwin, fifty years later, we have gone far beyond transformism, that is to say a general hypothesis positing evolution, as opposed to fixity, of species. His work represents the first true formalization of knowledge about living beings.

The essentials of Darwin's theory, based on a vast body of inquiry and observations and the fruit of deep and prolonged reflection, are well known. Individuals are engaged in a perpetual struggle for control of the environment; and this struggle confers advantages on some of the small variations that arise in them. Through hereditary transmission and accumulation of these favorable variations, species are gradually transformed. "This preservation of favourable individual differences and variations, and the destruction of those which are injurious, I have called Natural Selection, or the Survival of the Fittest," Darwin writes in *The Origin of Species* (1859). This theory combines into a coherent whole, through simple and accessible concepts, a considerable number of facts drawn from an extraordinary range of fields: biogeography, paleontology, comparative anatomy and physiology, morphology, embryology.

Darwin's ideas on natural selection are still fruitful and alive, and hence discussed, today. They have been updated by Ernst Mayr in the context of the theory of evolutionary synthesis, which offers a general explanation for the genetic diversity of populations; in part, they have recently been called in question by the neutralist theories (the Japanese M. Kimura) which treat the evolution of genetic structure as random, assigning only a secondary role to the mechanism of selection; and they have had new postulates added to them by the thermodynamicists, like Ilya Prigogine and Henri Atlan, who are advocates of self-regulatory systems.

But while Darwin's theory thus constituted a true scientific revolution, one whose results and insights we are far from having exhausted, we can in retrospect say that for molecular biology to become possible, at least two other lines of research needed to be developed.

For the theory of natural selection, while providing a first major principle for its rational organization, does indeed say nothing about the actual nature of the living world, in other words about what lies beneath the morphological and functional properties of living beings; it says nothing about the mechanisms of which they are made up.

Consequently, in order to tackle this formidable problem, biology needed to acquire greater analytical precision and invent other formal systems. To supplement Darwin's theory, researchers would have to delve deep into the interior of living matter. The cell theory was the first step in this direction. Of course, this theory did not just fall out of the blue one fine day. It has its own prehistory.

The Dutch naturalist Antonie van Leeuwenhoek (1632–1723) is regarded as one of the first scientists to have used the microscope to observe samples taken from living beings. Around the same time Robert Hooke, physicist, astronomer and naturalist, first described the cellular organization of a fragment of cork. Lastly, it is to Lazzaro Spallanzani, in the next century, that we owe the first descriptions of living cells, particularly spermatozoa. In truth, however, the cell remained at most an object of curiosity up until the beginning of the 19th century. It was only with the work of the German biologist Matthias Jakob Schleiden in 1838, and above all of his friend Theodor Schwann, that the cellular composition of plants, and then of animals, was categorically established. Schwann explicitly states that the cell must be regarded as the elementary unit of life. Other great biologists (Jean Evangelista Purkinje, Karl Wilhelm von Naegeli, Gottlieb Wilhelm Bischoff, etc.) confirmed these studies and elucidated some of the mechanisms of cell division. But it was Rudolf Virchow, a Prussian doctor and politician, who towards the middle of the 19th century took a decisive step forward by proclaiming in 1858 an aphorism which has remained famous: "*Omnis cellula e cellula,*" every cell comes from another cell.

As a measure of the revolutionary scope of this formula, it must be recalled that at the time of Lamarck, and even of Darwin, biologists still believed in spontaneous generation. They continued to think that lower organisms, infusoria for example, were under certain conditions created directly out of inert matter.

To be sure, this was no longer the age when the celebrated Flemish chemist Jan Baptist van Helmont (1577–1644), a disciple of Paracelsus, explained that leeches, slugs and frogs were created from the mud of swamps, and mice through "transmutation of a sack of wheat wrapped in a dirty shirt!" But it was not until the work of Louis Pasteur, whose *Mémoire sur la fermentation lactique* [Note on lactic fermentation] dates from 1857, that the idea of spontaneous generation was finally abandoned. And we know how great was the opposition Pasteur himself had to over-come, even in his time: the name of Archimède Pouchet has gone down in history on account of the rearguard action he fought.

With Pasteur and his pupils a new discipline, microbi-ology, championed by a French doctor, Casimir Davaine, began to flourish. It was not only etiology and the preven-tion of infectious diseases that were to be "revolutionized" by Pasteur's work; on the theoretical level, his achieve-ments were to promote the advent of modern biology, a biology based initially on the use of bacterial models, for which Pasteur had paved the way by focusing the interest of the scientific world on microbes.

Having glimpsed the unity of the biological world through cellular organization, scientists were henceforth to concern themselves with how a cell functions. The way was now open for them to take up the intimate chemistry of living matter.

Here again, as always, a long prehistory would have to be retraced in order to identify the antecedents of this idea. As far back as the 17th century, scientists had intuitively grasped that vital phenomena could be explained by the interaction of chemical forces. A name we have just men-tioned dominates this line of thought, the name of Paracelsus. But in his day chemistry was still no more than alchemy, and hence could only direct research down paths that led nowhere.

With Lavoisier (1743–1794) a change of course took place in the way the problem was posed. He was not content to confine himself, as is often taught, to introducing the systematic use of scales and to formulating the principles of conservation of mass and of chemical elements; he quickly came to understand the extraordinary importance of chemistry in the study of living beings. Who does not recall his famous aphorism: "respiration is combustion"? Moreover, his close interest in the functions which were later to be termed metabolic led him to discover hemoglobin, and to explain the origin of animal heat. The seeds of the concept of cellular energy were latent in his work.

But Lavoisier's work cannot be mentioned without referring also to that of his contemporary, the eminent English theologian and scientist, Joseph Priestley. After taking an interest in electricity, Priestley became passionately involved in chemistry, and in 1774 first produced oxygen by heating mercuric oxide; later, he discovered hydrochloric acid, nitrogen dioxide, etc. Like Lavoisier, he understood the importance of oxygen for vital phenomena. He noted, for example, that an animal placed in polluted air dies of asphyxia, and associated this phenomenon with the inability of such an atmosphere to sustain the combustion required for respiration. Lastly, it should be added that Priestley was the first to recognize the role of oxygen in plant respiration.

The way was now open to chemical analysis of the components making up the fabric and content of living beings. It became apparent that life is, chemically speaking, a basic combination of carbon, oxygen, nitrogen, sulfur and phosphorus—all of them being elements already discovered in inert matter and interacting in accordance with the same laws in inorganic substances and living bodies.

On the basis of this first great unifying principle equating living and inert matter, chemists and pharmacologists grew bolder and, twenty-three years after Lavoisier's death, dealt a decisive blow to animism, a theory which, as propounded by the German doctor and chemist Georg Ernst Stahl, attributed specific virtues to living matter and posited that the soul acts effectively on the muscles and the digestive tract and governs the development of the embryo. For the first time, with the work of Jean-François Derosne on narcotine in 1817, and of Joseph Pelletier on morphine and then on emetine and quinine, it was possible to relate the pharmaceutical virtues of plants specifically to the chemical properties of given substances, even though there was not yet any talk of molecules.

In 1828, Friedrich Wöhler first succeeded in synthesizing a substance present in living beings, urea. From then on, things moved very fast! Encouraged by their success, chemists were no longer content to explain the properties of living beings, but focused attention on the mechanisms at work in them.

Biochemistry may be said to have been born on the day, in 1897, when the German chemist Eduard Buchner succeeded in demonstrating the relationship between the action of yeast as a ferment and the presence in cellular bodies of a substance that could be extracted. This was the first enzyme to be discovered: zymase. That the importance of his discovery was quickly understood is shown by the fact that Buchner was one of the first to win the Nobel Prize for chemistry. From then on, biology was to be a science of enzymes.

Gabriel Bertrand, who trained my first teacher at the Pasteur Institute, Michel Machebœuf, and Richard Willstäter in Germany then launched an assault on the structure of these astonishing catalysts. Their work identified how enzymes act, and in particular developed the first ideas about "coenzymes," most frequently metallic in nature.

In 1926, James B. Sumner, an American biochemist, took this adventure a stage further by being the first to prepare an enzyme, urease, in the crystalline state. Another American, John H. Northrop, achieved the same feat for two enzymes long known for their activity in gastric juices, pepsin and trypsin. Sumner and Northrop were to share the Nobel Prize in 1946.

A new era in the life sciences thus began. This was the era I knew at the very beginning of my career as a researcher, the era of biological kinetics and thermodynamics, with the work of the German Michaelis and the Frenchman Charles Henry, who were the first to propose a theory, which still holds today, explaining the activity of these astonishingly specific catalysts, enzymes, the true molecular workmen of the cell.

I will not here go into the details of the research which, thanks to Koschland and then to Jacques Monod and Jean-Pierre Changeux in the 1970s, was to lead to the association with the concept of the "active site" present in an enzyme of the concept of its allosteric site. By occupying such a site, one can change the composition of the protein, and with it the affinity of the substrate for the enzyme, and hence also the speed of the reaction.

Looking back over this history with the necessary degree of detachment, one is struck by its highly remarkable continuity: it reflects a reductionist and analytical approach to living matter, inspired first by chemistry and then by physical chemistry. It is apparent, too, that this research led to a comprehensive and rational explanation of metabolism and energetics. Life is catalysis, with enzymes serving as the catalysts. And for any living cell, the great banker providing the energy is a phosphorus molecule whose chemical bonds harbor a very high potential for hydrolysis, that is to say for scission through the absorption of water: adenosine triphosphate (ATP). The chemical unity of living matter was thus clearly established, and must be seen as one of the main achievements of the new biology immediately following the Second World War.

When I came to the Pasteur Institute in 1945, to the Biochemistry Department, with the firm intention of going into biology, there was no biology that was not chemistry. Obviously, I was not aware at the time that the true revolution in the life sciences was still to come. Initially, the way to it was to be paved by the intersection between the line of research I have just described and another such line relating to the laws of heredity. Once again, we need to turn to history in order to understand the significance of this moment.

The first serious experimental attempts to produce crosses and observe their consequences probably date back to the Frenchman Pierre-Louis de Maupertuis, a brilliant mind who excelled in all branches of knowledge: mathematics, physics, literature and natural history. It would however undoubtedly be wrong to make Maupertuis into a precursor of Mendel on the grounds that he proposed a particulate theory of heredity. His theoretical framework was so different from that of modern genetics that there can be no justification for establishing a link between them.

It was with Charles-Victor Naudin, a botanist born in 1815, that the study of hybrid plants was to be systematically resumed. At almost the same time as Mendel, he proclaimed the law of the purity of gametes which we shall be discussing below, and in this light can be regarded as one of the founders of genetics.

Properly speaking, this science began only with the publication by the Brno monk Gregor Mendel, in 1866, of the results of his patient statistical studies on crossing peas. Mendel did not content himself with the long-accepted idea that the characteristics of the parents were combined or mixed (Darwin) in the offspring; he studied the development of the characteristics taken individually. Organizing his experiments on a broad scale from 1854 to 1863; he studied almost 28,000 plants of the *Pisum* (pea) genus.

The first important discovery was that the individual characteristics of plants do not mix in hybrids, but are transmitted as discrete quantities. This led him to continue his experiments with succeeding generations, and to conclude that plants, like other sexual organisms, reproduce by means of gametes: the male pollen and the female ovum. In the process of reproduction, the total number of genes (a term not used by Mendel) provided by each of the two parents is halved, and each gamete contains only a single factor.

In addition, Mendel presented the results of his experiments in mathematical form, and was thus able to write "The constant characters found in different forms of a group of plants can, through repeated artificial fertilization, yield all the groupings indicated by the law of combinations."

Although they had direct and immediate knowledge of them, the best experts of the age did not understand these studies. It was only at the beginning of the present century that Mendel's laws were rediscovered and their general applicability and relevance to both the animal and the plant kingdom confirmed. They were to undergo a genuine resurrection at the hands of thinkers such as William Bateson in England and Lucien Cuénot in France.

At this time, the phenomena of meiosis and mitosis (cell division with or without reduction of the number of chromosomes) were known. The German Walther Flemming and the Belgian Edouard van Beneden had defined their cytological characteristics, that is, those inherent in the cell. The nuclear material, the carrier of heredity, was baptized chromosome by Wilhelm Waldeyer in 1888. Genes were to be christened by the Dane Wilhelm Johannsen in 1909.

Lastly, a German biologist, August Weismann, known as the father of neo-Darwinism and celebrated for his opposition to the Lamarckian theory of inheritance of acquired characteristics, put forward the hypothesis of germ plasm, in works published between 1892 and 1902. While the terminology has an obsolete ring today, the substance of what he said remains true: the nucleus of the germinal cell contains the hereditary substance carried on the chromosomes.

Hugo De Vries, on the basis of studies on pigmentation changes in a test plant (*Oenothera*), simultaneously introduced the concept of mutation, a sudden variation affecting the hereditary material (1900). In the United States, Thomas Morgan (1866–1945) and his collaborator Hermann Muller, experimenting on an extremely favorable material—the vinegar fly or *Drosophila melanogaster*, later commonly known as *Drosophila*—, evolved the chromosome theory of heredity between 1910 and 1920.

The genes, carried by the chromosomes and arranged, in their view, like a string of beads, are the true hereditary units. They represent simultaneously the unity of variation (de Vries) and of transmission (Mendel). They can undergo rearrangements by "crossing over" in the course of the pairing of the parental chromosomes that follows fertilization. Thus an individual's characteristics can be mapped by determining the relative positions of his genes on the chromosomes. This is what is known as genetic topology (from the Greek *topos*, place), knowledge of the distribution of gene locations.

Genetics was born. Its data and its theories were to fire the minds of scientists for generations, right up to the present day. But above all, as we shall see, with these new concepts and those which had emerged from the study of the cell and of enzymes, a third major unifying principle was to be revealed.

A new history was beginning, no less complex than those which preceded it.

THE EMERGENCE
OF MOLECULAR BIOLOGY

Shortly after the end of the Second World War, it could be considered that, conceptually speaking, biology had, one might say, run its course. The logic of life, it was thought, had in its essentials been definitively revealed. Had it not been firmly established that the genes, the cornerstones of modern biology, controlled the formation of enzymes, those all-purpose proteins on which the main functions of living matter depend? Had not Boris Ephrussi, George Beadle and Edward Tatum, through their remarkable studies—the first on *Drosophila* and the other two on *Neurospora*, a yeast—succeeded in bringing together genetics and biochemistry by showing that each gene corresponds to a given protein chain? The famous equation then triumphantly put forward—"one gene, one enzyme"—is well known. In short, biologists were convinced that, now that the functioning of the genes was known, the cell had yielded up all its secrets.

Nevertheless, certain questions remained unanswered, and they were not minor ones. What are genes themselves made of? What, precisely, is their physico-chemical structure, and by what strange magic is the transition made from a gene, which gives the orders, to the protein, the craftsman of cell chemistry? Should one not

31

imagine a specific machinery, housed in the cell, capable of both understanding and reading the gene's code? And, lastly, what about the code itself? These were highly embarrassing questions, which for years haunted the theoretical consciousness of biologists.

Indeed, in order to answer them adequately, an obstacle had to be removed: the belief then almost universally accepted, strange as it may seem today, that genes *were* proteins. The astonishing performance of enzymes was recalled, and the conviction prevailed that cell components other than proteins were involved at best only as supporting elements.

Once this obstacle had been removed, the extremely curious substance that a young Swiss chemist, Johann Friedrich Miescher, had succeeded in isolating in 1869, the same year Mendel submitted his famous second paper on plant hybrids, was rediscovered. This nuclein, the substance of the nucleus, abundantly present in leukocytes (the blood corpuscles which defend the organism against pathogenic agents) and in salmon roe, was none other than deoxyribonucleic acid, DNA for short, which received its name only after its chemical structure had been determined.

I will not recapitulate here the body of research which led to the abandonment of the idea that genes were composed exclusively of proteins, and then to the identification of DNA, discovered by Miescher, as the key molecule through which all cellular life perpetuates itself. I have already recounted the central events of this exciting story in my book *Les Secrets du gène* [The secrets of the gene]

(Odile Jacob, 1986). Let me simply recall, as a reminder, the decisive work of the American pathologist Oswald T. Avery on the transforming principle in pneumococcus (1944) and that of Hershey Chase on the mode of transfer of genetic information in bacteriophages (1952).

On the other hand, it does seem to me useful, before reviewing the answers that investigation of the properties of DNA offers to the questions that had remained pending, to draw attention to an important epistemological aspect of the birth of molecular biology.

The role, in many respects decisive, played in the history of the life sciences by the exact sciences is well known; whether through mechanics with Descartes, chemistry with Lavoisier and Pasteur, or statistics with Mendel, they not only exerted a magnetic attraction on the life sciences, but provided them with a model. From 1950 onwards, it was the nuclear physicists and crystallographers who took center stage and, it is no exaggeration to say, shaped the new face of biology—the face that we know today. This was the time—is any reminder needed?—when information theory was flourishing and electronics was beginning to make its mark: William Shockley discovered the transistor, and the first computers appeared. An Austrian physicist, Erwin Schrödinger, well known for his masterly contributions to wave mechanics, the principles of which had been established by Louis de Broglie, began to take an interest in the phenomena of life. In 1945, he published a work which immediately caused a sensation. Under the title *What is Life?*, he discussed the key questions raised by genetics, and established a conceptual link between information theory and the knowledge then available of the cell macromolecules. He may be said to have laid the foundations for molecular biology.

In 1952, James D. Watson and Francis Crick came up with the first partial answer to the fundamental question posed by Erwin Schrödinger seven years earlier. Life is explained by genes, and genes are found on a very long molecule in the form of a double spiral, DNA, coiled up within every cell. Maurice Wilkins very soon established its structure by means of x-rays. Thus emerged the celebrated model of the double helix, caricatured as often as it has been faithfully reproduced, its admirable esthetics celebrated by Salvador Dali himself, in his own fashion. This structure, consisting of two parallel complementary chains linked together by easily dissociable bonds, affords a remarkable explanation of the transmissibility of characters whenever a cell divides: once separated from the other, each chain of the helix begins, like a crystal, to form two double helices identical in structure to the original.

This model lies at the very heart of the reborn biology. All our knowledge of the intimate mechanisms by which genetic material is reproduced derives from it, as does the knowledge we have gained about the transfer of information which takes place in protein manufacture. The remarkable book by James D. Watson entitled *The Double Helix* (1968) gives an insight which is as lively as it is objective into the approach taken by those whose names have now become inseparable from the history of our science.

One of the key stages in this astonishing odyssey, covering the period between the mid-1950s and the mid-1960s was to be the discovery of the intimate code of the cell, by which I mean the genetic code.

The brilliant insights of scientists like Francis Crick and George Gamow were extended experimentally through the work of Marshall Nirenberg, and resulted in one of the most spectacular achievements of modern science: the complete decoding of the sixty-four combinations of nucleotides, that is to say of the links in the nucleic acid chain (also known as codons or triplets because they consist of three juxtaposed elements, the nucleotides, each of which is composed of one base, one sugar and one phosphorus atom).

The analysis (one might say dissection) of the biochemical stages through which this three-letter sixty-four-combination code is transposed into a single-letter twenty-combination code constitutes a triumph of modern biology. It was made possible through the discovery, in 1961, of messenger RNAs (François Jacob, Jacques Monod, François Gros), ephemeral copies of the genes which provide an assembly template for protein synthesis.

However, it would take no less than fifteen years of arduous labor to elucidate, bit by bit, all the components of the surprising machinery which allows this assembly to take place, at a rate of five hundred links a minute, and with a negligible error rate.

The astonishing analytical power of molecular biology is demonstrated by the discovery of ribosomes (RNA-rich particles whose properties had already been studied by Jean

Brachet and Torbjörn Caspersson in 1950), of transfer RNA (adaptor molecules which arrange each amino acid to line up with each specific triplet of the messenger), and of activator enzymes and the countless translation factors (effecting the transition from the code to the chemical mechanism).

There remained a last question—that of the influence of the environment on genes. Indeed, the prevailing belief at the time was that acquired characters were not transmissible. The organism adapts, it develops new functions through a learning process, but what is acquired never becomes innate. It does not pass into the genes. For the descendants to inherit it, they must be located in the same environmental context as their parents and the properties of their genes (the famous genotype) must so permit.

But if the structure of the genes no longer retains the imprint of the many physiological or intellectual gains made by the individual, this in no way means that the environment, understood in the sense of experience, cannot temporarily modify the functioning of these genes. Jacob and Monod, thanks to the discovery of regulator genes, were the first to be able to shed light on gene cybernetics: whereas most genes control cell activity or morphology through proteins whose synthesis they regulate, there exists within the cellular factory a remarkable and extremely logical hierarchy. The regulator genes, through the proteins they manufacture, modify in one direction or another (activate or repress) the activity of the structural genes.

Twenty years or so ago, therefore, biologists had every reason to feel satisfaction. The formal precision the life sciences had hitherto lacked was beginning to emerge, with the central dogma DNA $\rightarrow$ RNA $\rightarrow$ protein, and with the discovery of the code and the regulatory circuits. Biology, initially a science of inventory and classification in the hands of the 19th-century comparative naturalists, later transformed with the geneticists into a statistical science and then, with the prewar biochemists, into a powerful analytical discipline, had become a science of codes and circuits, thus taking on a strange resemblance to microelectronics and information science. Biologists and physicists were now able to understand one another, subscribe to similar objectives and share the same enthusiasms for a Universe which was apparently beginning to yield up its secrets.

And yet this period, marked by the appearance of Jacques Monod's book *Chance and Necessity* (1970), was to go down in history as the end of a reign rather than as a triumphal stage in the progress of a science on the march.

Was this because of the scientific fatalism professed by Monod himself in this well-known book, in which he calls on his contemporaries to celebrate with him the virtues of an austere "ethics of knowledge?" Remember the sentence: "Science disregards values: the concept of the Universe it imposes today is devoid of all ethics." Or was it that the discipline was out of breath, and its protagonists needed to pause for a moment to reflect better on what had been achieved?

The fact remains that biology and biologists had, with their remarkable molecular advances, somewhat cut themselves off from the world. Their ignorance of some of the social and economic realities on which biology was to have a profound impact cannot fail to strike the observer. This was the time when a scientist like Erwin Chargaff went so far as to proclaim, with some bitterness, that molecular biology had never done anything to advance medicine. A modern-day Diogenes, he heaped sarcasms on the arrogance of the new adepts of a science whose splendid isolation he held up to ridicule. The only advances made by medicine stemmed, in his view, from the use of antibiotics and steroids, the outcome of successful research by the biochemists of an earlier day. Chargaff may perhaps have too easily given free rein to what might pass for personal rancor; nevertheless, he said out loud what many at that time were thinking in private without daring to express it.

It must be admitted that the biology of that time, with its somewhat dry models more likely to satisfy rigorous mathematicians than naturalists, with its circuits, dogmas and codes, did indeed strike most people as cold and esoteric. True, its results were seized upon by the most formalistic of the linguists, seeking through its authority to establish their discipline as a pilot science in the field of the human sciences; and at the same time, the torch had been taken up by anthropology in its "structuralist phase." A whole combination of circumstances was involved to which ultimately Jacques Lacan, combining the lessons of the linguist Roman Jakobson and the anthropologist Claude Lévi-Strauss in support of the scientific modernity of his rereading of Freud, had brought all the brilliance of his rhetoric and the then growing intellectual and social acceptance of psychoanalysis.

By reducing life to molecular interactions, had not biology, despite the victories it had won, abandoned the big questions? What had it done for people, our health, our well-being and also, since the ecological movement was then in full swing, our environment? These were the burning questions of the day.

Are there grounds for speaking of a "crisis of confidence" specifically affecting the life sciences? And to what extent should this moment of wavering be associated with the broader conflict of the challenge to the scientific and technical approach which was to find full and virulent expression in May 1968? The question remains open. What is certain is that from then onwards the concepts and methods of molecular biology were to be applied to fields which were external to genetics and had long been neglected. This was particularly true with regard to studies of the brain and the nervous system.

THE MOVE UP TO
HIGHER ORGANISMS

The investigation of neurons, their activities and the relationships between them was still almost exclusively the province of the neuroanatomists and neurophysiologists, even though at a higher integrative level psychophysiologists and psychiatrists were also involved. Molecular biologists had not yet dared to attack what was for them *terra incognita.*

Günther Stent did not hesitate to write and proclaim that the golden age of molecular biology (the age of the gene) was over. Interest was shifting, he announced, to the neurosciences. And indeed, many molecular biologists, and not the least outstanding among them, decided to become neurobiologists: Marshall Nirenberg, the discoverer of the code; S. Benzer, who revealed the fine genetics of bacteriophage; and, a little later, Francis Crick himself.

This extremely broad and remarkably synchronized movement was to give birth, from 1978 onwards, to what some have called molecular neurobiology—the study of the major components of the neuron by means of genetic engineering and monoclonal antibodies.

Other molecular biologists, fearing the difficulties of the neuroscience field, nevertheless preferred to move on to the study of higher organisms. *Escherichia coli*, so called because it was discovered by Theodor Escherich and is found in the colon, is indeed a valuable bacterium, part of our natural fauna, an extremely simple organism, without a nucleus, that has a relatively small genome and a very rapid division rate, thus offering ideal material for genetic analysis. Monod argued in vain that "everything which is true of *Escherichia coli* is true of the elephant"; there were many who, at the end of the 1960s, wished to attack the mechanism of cell differentiation and the development of eukaryotes—the cells of higher organisms which, unlike prokaryotes (bacteria, algae, etc.) have a clearly defined nucleus enclosed within a special membrane. François Jacob was one of these; Sydney Brenner was another. After a number of tentative efforts, the former took up the development of mice, while the latter focused his interest on the nematode, a helminth or tiny worm which nevertheless has a complex nervous system and a true musculature.

Jacob's ambition was to be able to make a significant approach to the first phases of the development of organized beings—and thus to go further than the traditional embryologists had done by explaining the molecular and cellular algorithms of the beginnings of embryogenesis. This was a formidable challenge at a time when only the egg of the batrachians (frogs, *Xenopus*) lent itself to quantitative embryological study. Jacob, and others with him, thought it would be more profitable to work on mammals. Mice were already known to lend themselves well to genetic studies, in that they were found to undergo a large number of identified and catalogued mutations. It was also known, however, that the mouse egg was small, and hence somewhat resistant to investigation by the techniques of biochemistry and molecular biology. Fortunately, Beatrice Mintz, Boris Ephrussi and a few others demonstrated the existence in this small mammal of tumors of the germinal tissue, known as embryonic carcinomas or teratocarcinomas, which show properties similar to those of normal blastocytes, but with the valuable difference that they can be cultured and lend themselves to biochemical analysis. Thus it was discovered that in some of them, differentiation into a great variety of tissues can be induced *in vitro*. Hence these murine teratocarcinomas would become the preferred line of research of the Pasteur Institute, following the course set by François Jacob.

For S. Brenner at Cambridge, as we have said, it was nematodes, and for S. Benzer at Purdue University (Indiana), *Drosophila*. Some neurobiologists preferred *Gymnotus*, a variety of electric ray in which acetylcholine receptors are particularly abundant. Acetylcholine, a mediator in cell excitation, is found throughout the animal kingdom, and in vertebrates in all the motor and sensor nerves, the synapses, etc. Still others chose *Aplysia*, a marine snail frequently found on the Atlantic coasts. Lastly, there were those who, like me, decided to concentrate on the muscles in large and small mammals, not to mention those who opted for the red blood cell of the duck or humans.

If any general lesson can be drawn from this hunt for models, it is perhaps that it reveals an approach which is characteristic of the experimental sciences: when researchers are short of concepts, they change models, becoming naturalists again, as if seeking to build up a new stock of unknown material.

However that may be, for several groups of reasons whose relative importance is difficult to assess, molecular biology was soon back at the focus of social concerns. I would single out two of these reasons, which appear to me to be the most important. The first stems from the reaction of the pharmaceutical industry, in full expansion since the postwar period, which when faced with sudden pressure from the environmentalists became more aware of the need to use cellular or molecular models to inform the public of the positive, harmful or iatrogenic (disease-causing) effects of the products being marketed.

The second reason stems from developments in medicine and from the restructuring of medical studies. Research disciplines which had hitherto been considered the almost exclusive province of doctors, or at best of pathologists, began to draw their inspiration from molecular biology. Examples are virology, immunology, cancerology and neurobiology.

Were these moves initiated by doctors or by molecular biologists? Was there not, rather, a convergence of interests, not all of them strictly intellectual in nature? This was recently suggested by Lewis Thomas, former president of the famous Sloan-Kettering Cancer Center in New York and regarded in the United States as one of the most lucid minds where the philosophy of science is concerned:

"I do not remember hearing the compound word *bio-medical* when I was a student. I believe it was introduced shortly after the Second World War, concocted, I imagine, in accordance with the rules of the jargon by tacit agreement between the two communities in the hope that it would yield them the benefit of public recognition. Doctors doing research must have welcomed the prefix *bio* because of the prestige it conferred on the profession; conversely, biologists proper must have felt some need for the suffix *medical*, perhaps to help them obtain government grants."

The fact that it is expressed in blunt terms is no reason for rejecting the substance of this diagnosis. The life sciences undeniably became "organismic" again—they returned to organisms as the subjects of study. Their clearly stated research objectives bear witness to this: the mechanisms of virulence, socio-occupational diseases, hereditary anomalies, cancer, behavioral disturbances and, at a still broader level, human reproduction and aging. In addition, spurred on by the anticonsumer movements, biologists at the same time also began to take an interest in ecosystems.

But none of this would suffice to explain the renewed popularity the life sciences are now enjoying or, conversely, the rejection reflex they sometimes trigger. This strong ambivalence of public opinion towards the life sciences is quite certainly a result of the emergence of genetic engineering and its repercussions for biotechnology.

THE 1972 TURNING POINT

It will perhaps be recalled that the beginnings of this decisive new turning point in molecular biology became apparent in 1972. That was the year of the first publications dealing with artificial recombination techniques which offered the possibility of cloning—reproduction of matching copies—of genes as well as their large-scale purification, or their transfer into cells by pathways which defeated natural barriers.

With the advent of these bold techniques the public rediscovered the life sciences, perhaps for the first time since the epic achievements of Pasteur. By the same token, it recognized that they could be disturbing, just as the atomic sciences could. At the same time, the world of decision-makers, politicians and industrialists understood that a modern technology, one which was both a motive force and a product of the biological sciences, had been born, from which major, indeed revolutionary, practical consequences could be expected in the fields of public health, agriculture, animal husbandry, energy production, chemistry and the environment. Biology was becoming, very clearly, a hands-on science. Nor did it take long to discover that it could even become a business.

I have emphasized the turning point which drove research on again to new horizons because it appears to me fundamental to an understanding of the recent achievements of molecular biology and their most topical social impacts. It should not, however, be concluded from this that the bacterial model was henceforth consigned to oblivion, relegated to some remote corner of a history now over and done with, and that it had thus purely and simply given way to models based on more elaborate organisms. Research never follows such a course. What happened was very different: microorganisms in general, and *Escherichia coli* in particular, were part of the same movement subjected to a second and even more detailed exploration—"revisited," so to speak. As was to be expected, it emerged that the differences between the cells of these microorganisms and those of eukaryotes are less marked than had been thought. For example, the genetic regulatory elements found in them have certain points in common with those of higher organisms.

J. Yaniv has even described in *Escherichia coli* proteins which bear an extremely curious resemblance to those found associated with the DNA in the chromosomes of plants and animals: the histones. In addition, certain processes associated with the transport of bacterial proteins from their site of formation to the environment have been elucidated (M. Schwartz), and these, too, show characteristics similar to those observed in protein secretion in eukaryotes (C. Blobel).

The fact nevertheless remains that, alongside these advances in the study of the microbial cell, research also moved on to a new stage, in which physiological phenomena of growing complexity were tackled.

A few examples will illustrate this. Let us take, first, the study of gene cybernetics in large organisms.

Among bacteria, the phenomena of regulation by genes were essentially explained in the famous operon model proposed by Jacob and Monod in 1960. This model implies, it will be recalled, the existence of topologically grouped genes which are thus controlled as a unit by the interplay of repressor or activator proteins acting upon specific operative sequences.

In higher organisms, the studies made possible by genetic engineering reveal the existence of infinitely more complex mechanisms, even though certain general principles, such as the interaction of a signal protein with its operator, remain valid. Thus a great diversity of target genetic sequences is becoming apparent close to the site at which recognition of the gene by the transcriptases—the enzymes which catalyze the transcription of the genes into RNA as the first phase in genetic expression—begins. These target sequences are receptors for a large number of regulatory proteins. Some of these proteins are common to the majority of genes, others are specific to certain genes, and still others go into action only when the gene is operating in a specific tissue context.

The whole range of regulatory proteins and transcriptases can form three-dimensional assemblies, thus imprinting on the genetic material spatial configurations whose laws are beginning to be deciphered. A full-scale physical chemistry of the protein-DNA interaction is thus emerging. It is apparent that it involves discrete regions on the protein chains known as regulatory areas, shaped like the fingers of a glove.

However, it seems that the control of genetic activity must involve another essential factor, namely the chemical changes taking place in the regulatory proteins themselves. Produced by specific enzymes which alter the properties of the proteins, these changes, known as post-translational, consist for the most part of phosphorylation triggered by specific kinases.

In the field of the general biology of eukaryotes, another major advance lies in the discovery of the mechanisms which transmit an external signal (hormone, neurotransmitter, growth factor) from the membrane to the boundary of the nucleus. These phenomena are often termed transduction. This is an extremely important discovery in that it affords an explanation of many hitherto mysterious aspects of normal cell homeostasis, and also, as we shall see, of the cancerous state. The term homeostasis is used to denote the state of a system which compensates through a series of reactions for the variations taking place in it, and tends to return to the initial state.

Within the space of a few years, the sequence of events triggered when a cell is activated by a factor whose function is to stimulate its growth had been identified.

Most of the biochemical reactions, and of the molecular intermediaries (often termed second messengers), had been ascertained.

These studies have led to the discovery of a new cycle of intracellular changes which makes it possible to account for the smallest details of how a plant or animal cell obeys the chemical signals emitted by other cells. With this discovery, molecular biology touches upon the formidable problem of cellular communication.

Over the past ten years, we have been beginning to arrive at a better understanding of phenomena as complex as intracellular motility, secretion, contraction, vectoral migrations, and the phenomena of adhesion and fusion, or again of deformation, which form part of what is known as biomechanics. The physicochemical properties of the majority of cell microstructures, and of the guiding molecules (fibronectins) or the binding molecules (lectin), are beginning to be known. As these data accumulate, a more detailed picture is thus emerging at the molecular level of what gives the cells of higher organisms their astonishing powers of differentiation, tissue assembly, communication and movement. But as we move higher up the scale of complexity of life, we find ourselves faced with other problems of the architecture and composition of cell groups when we seek to explain the spatial geometry of an embryo or of the individual into which it grows.

Embryologists and cell biologists had long been concerned with the overall morphogenesis of living beings, investigating the nature of the information that dictates to the tissues, organs or limbs their respective positions in space as well as their dimensions. Quite recently, this particularly sensitive field of research has benefited from surprising discoveries. A new class of genes has been identified, which could be termed architectural genes.

It is now known that most embryos, from the fly to humans, are built up as if with an Erector Set: they consist of assemblies of segments which, for a given species, are finite in number and each have a specific purpose; for example, the thoracic and abdominal segments of insects which will develop into legs, wings, head or antennae, or the mesodermal segments, or somites, in mammals, those segments of the layer in the embryo known as the mesoderm which are destined to form the principal muscle masses. Each of these modules has its own autonomous genetic determinism. Under certain circumstances, for example following certain mutations, it can evolve on its own, independently of the overall architecture, and this will lead to the formation of abnormal individuals.

Such monstrosities have been known to experts for a very long time: *Drosophila* with a double thorax, two pairs of wings, or antennae in the place of legs. Thus, it is known today that specific genes determine the number and polarity of the segments; others, the homeotic genes, shape their development. Generally, the course of embryogenesis is normal, because these homeotic genes communicate with one another and a precise balance is established in their operation. But if this communication and this balance are absent, a disturbance may occur which opens the way to teratogenesis, the creation of monsters.

MOLECULAR BIOLOGY AND CANCER

For the first time, then, a bridge has been built linking the generally descriptive science of embryology to genetics and molecular biology. Pathology, too, is entering upon a new era thanks to the recent progress in biology. The most spectacular of these advances have undoubtedly been achieved in that key field of human biology and medicine: the study of cancer.

It has been known since 1914, from the work of American researchers, that some forms of animal cancers were attributable to viruses. In 1975, D. Stehelin, J.M. Bishop and H.E. Varmus discovered a specific new category of genes—oncogenic cell sequences or oncogenes for short. It has been established today that many human cancers result from the action of viruses known to be oncogenic (from the Greek *onkos,* mass or bulk).

Such is the case with the hepatitis B virus, frequently associated with primary cancer of the liver, or with the Epstein-Barr virus, responsible for cancers of the jaw in Africa and Asia, or again with the papillomaviruses which cause cervical cancers. In 1983, another category of viruses was discovered containing RNA instead of DNA, the retroviruses, which can cause acute leukemia (HTLV 1 and 2); the various classes of viruses responsible for AIDS (HIV 1, 2 and 3) are also known to be capable of giving rise to certain cancers (Kaposi's sarcoma). And it has been established that there is a link between the carcinogenic capacity of these viruses and the presence in their hereditary material of a specific gene: the viral oncogenic sequence.

The surprising discovery has been made that all the cells of eukaryotes include, *even in the normal state*, gene elements which are very similar to these cancer genes that had hitherto been described only in viruses. For a time, the existence of these cell oncogenes appeared to be a genuine paradox, all the more so in that they were ultimately also discovered in lower eukaryotes, such as yeast.

We now know that with this information we undoubtedly hold in our hands, for the first time, the genetic explanation of the onset of cancer. In fact, these cell oncogenes are none other than the genes responsible for the chemical intermediaries which govern cellular communication. They thus play an important role in the economy of the normal cell. When these determinants undergo any mutations, expansions or rearrangements under the influence of environmental factors, the cell suddenly finds all its homeostasis mechanisms disrupted. It ceases to obey the growth factors, initiates anarchic division, and becomes malignant.

It now appears that the carcinogenic viruses owe their properties to the fact that, very early on, at the dawn of evolution, they stole normal oncogenic sequences from the cells they infected. Once integrated into the genome of the viruses, these sequences acquired new properties, making them unresponsive to any commands by the cell.

This discovery of oncogenes obviously constitutes a major achievement in fundamental pathology. However, it is not only of great heuristic value, but could open the way to new forms of treatment, especially since genes have recently been discovered whose activity seems to be aimed at blocking the cell oncogenes and which could mitigate the disruptions they cause: the genes which code for beta-TGF (*tumor growth factor*), TNF (*tumor necrosis factor*), and gamma interferon. Is it possible that through this discovery we have taken what might well be the first step towards a rational therapy of cancer? This is admittedly not the first occasion on which researchers have been inclined to raise questions of this kind. But never before had such precise information been available on the subtle struggle between genes and molecules, a struggle whose outcome determines the harmony or death of the cell!

A whole book would have to be written to give a full picture of the new breakthroughs made by molecular biology in its exploratory approach to the major functions of living beings.

Mention would have to be made, for example, of the remarkable progress being achieved in such disciplines as virology. The elucidation of the entire structure of the AIDS virus by Luc Montagnier and Robert Gallo is an exploit which is on everyone's mind today. The same is true of immunology, with the discovery of the lymphokines (the chemical mediators of immunity), the elucidation of the molecular structure of antibodies, the detection of histocompatibility genes, which regulate the individual's ability to recognize exogenous cellular or viral elements as forming or not forming part of the "self." However, we shall conclude this brief inventory with the achievements that fall within the field of the neurosciences.

MOLECULAR BIOLOGY
AND THE NEUROSCIENCES

By offering the possibility of analysis of greater precision than ever before achieved in studies of the key elements of neural communication, genetics has lent a new dimension to the sciences dealing with the nervous system. We have seen how, back in 1970, molecular biologists had decided to undertake the conquest of the neuron. This ambitious goal did not fail to give rise to a certain degree of legitimate distrust on the part of neurobiologists. How could one hope, by studying the genes active in neurons, to be able to grasp the awesome complexity of the neuronal networks, their development and their operation?

Moreover, the importance of the first results produced by molecular biology must not be exaggerated; they have no meaning unless they are situated in a context which only the classic disciplines of the neurosciences are capable of defining.

By the beginning of the 1970s the work on the chemical mediators involved in synaptic transmission, in the contact zone in which neuron-to-neuron relationships are established, followed by analysis of the subtle morphology of the neuron, its extensions and terminations, its sheaths, etc., afforded an understanding of the essentials of synaptic function. The entry of molecular biologists into this field made it possible to define the nature of the interactions between these neurotransmitters and their receptors when they leave the synapse. Above all, it resulted in more exact knowledge of the proteins of the neuron cell, from the moment when cloning of that cell's genes became possible. Remarkably precise information is now available on the organization of receptors on the surface of neurons, and particularly of the most important among them, the cholinergic receptor (cf. the work of Jean-Pierre Changeux).

This is certainly not the place to describe in detail how the discovery of the neuropeptides—and particularly of the endorphins or pain peptides in the 1980s—opened up a new era in the study of brain chemistry and interneuron communication.

Nevertheless, it is of interest to recall that the history of the neuropeptides has its origins in the story of opium, a substance known for at least six thousand years. The Sumerians already used it four thousand years before our era! Its analgesic and antidepressant properties were known to the Greeks. It was above all the Arab merchants who promoted its widespread production, by introducing it into China under the Tang dynasty. Only fairly recently, however, have the effects of this substance and its derivatives on psychology and behavior been the object of intensive research among the community of pharmacologists, biologists and psychologists, and this has been in very large measure since the discovery, in animals belonging to widely differing phyla, of a particular class of peptides, generally known as endogenous opioids because their action on the nervous system mimics that of opium proper.

Today, with the help of genetic engineering, the list of these neuropeptides has become considerably longer, and the keyboard of neuron interactions has been enriched by knowledge of almost thirty such molecules. Their release from the neuron cell and their reception by the target neurons offer an explanation of many forms of behavior. The work of the teams led by E. Kandel and R. Axel on *Aplysia* neurons afford a striking illustration of this.

While the molecular biology of the gene in its application to neurons is as yet for the most part only at the initial inventory stage, there is a clear feeling that the nervous system will gradually become amenable to increasingly subtle analysis. After study of the neuron in general, this will make it possible gradually to approach the functioning of integrated groups. A genuine molecular neurobiology, also supported, as we shall see, by studies of the genetic determinism of neuropathic conditions, is thus coming into being.

II

THE BIOTECHNOLOGICAL
REVOLUTION

The gigantic leap forward in biology which we have just
briefly described may in many respects be compared with
that which has taken place in physics and electronics in
the course of recent decades. It came about, as we have
seen, through the establishment of closer ties between the
life sciences and other sciences such as mathematics,
physics and chemistry. Will it result in the creation of new
frontiers, areas in which our knowledge will expand even
more rapidly? This indeed is what Ilya Prigogine insistently suggests. In his view, a certain convergence which
is apparent between all the sciences dealing with complex
networks—artificial intelligence, information theory, dissi-
pative systems, chaotic situations, etc.—and biology itself
could open up unsuspected prospects for an understanding
of the apparently most irreducible phenomena of living
matter—the organization and functioning of the cortex, the
cerebral representation of forms and neurobehavioral patterns—as well as for the analysis of ecosystems and
population trends.

These views bear witness to a powerful aspiration to
unify science, and they have given rise to extremely interesting intellectual debate and to original and stimulating
work; but perhaps the most significant indicator of biology's move to the forefront of the social scene needs to be

sought elsewhere. It is to be found, rather, in the cultural and economic spin-off from the already considerable achievements of molecular biology.

Are we soon to attain a degree of precision in describing the machinery of the cell and the individual which will give us real mastery over our own fate and that of the biological world in general? It is obvious that even now, with the techniques of recombinant DNA, cell engineering and transgenosis, today's biology has already acquired considerable capacity for intervening in all forms of life and affecting people as well as microorganisms, even though it may not yet be said to have mastered speciation itself, that is to say the creation of new species.

Is not biology able to help articulate and answer the questions being raised by the environmental movement, a movement whose initial inspiration derived from a reaction against scientific reductionism, which it saw as inseparable from the modern world's consumerism? Whether we welcome or deplore it, respect for the environment is now most frequently associated with the imitation of biological nature.

The new goods and services that have become available as a result of the most recent developments in molecular biology, in the field of public health as well as in those of agriculture and energy production, engender both hopes and fears. Some of the most vital questions of our time relate to how the social and human benefits of these developments are to be appraised. Inevitably, hope is aroused that these goods and services will contribute to solving some of the most serious geoeconomic and geopolitical questions. Now that, thanks to genetic engineering, new procedures are available for modifying the properties of microorganisms, plants and even animals, now that made-to-order plant forms can be reproduced on a very broad scale, there is a temptation to see in these abilities a solution to the economic difficulties facing the developing countries. Even though the speed with which it is sometimes forecast that the progress made in plant biology or animal husbandry will be transferred to these countries in concrete terms may justifiably be doubted, and even bearing in mind the fact that these applications are not without their harmful effects on the natural products market and the local labor force, there seem to be some grounds for this hope, and international organizations have embarked upon a process of thought and reflection that may in time lead to a better future.

Clearly, the life sciences, whose influence now extends to all aspects of human destiny, are being appealed to and challenged from all sides. Their cultural, social and political impact is extremely far-reaching. There is justification, I think, for speaking of a true gene civilization coming into being. The acuteness and urgency of these questions being posed by society can be demonstrated by describing a few characteristic situations.

The first topic that comes to mind when these questions are mentioned is the field of biotechnologies.

In 1978, together with François Jacob and Pierre Royer, I had the privilege of being commissioned to write a report for the French President on the life sciences.

The aim was "to study the consequences the discoveries of modern biology may have for the organization and functioning of society, to identify the applications of bio-technologies offering the greatest benefits for human progress and happiness, and to propose suitable means of implementing these applications" (letter from the French President commissioning the report on *Life Sciences and Society*).

Valéry Giscard d'Estaing, returning from a journey to Brazil, had been struck by the progress made in the field of bio-organic fuel production. No doubt this visit must be seen as the decisive factor which aroused his interest in biotechnologies and the life sciences. The report we presented to him at the Elysée is somewhat out of date in its descriptive section, the part relating to applications. Thus, on rereading the chapter dealing with the spin-off from genetic engineering, one is struck by the fantastic progress made in the past ten years. Joint ventures were then being hailed as a great innovation. We also come across sentences like the following: "Genetic engineering technologies are still not highly developed in countries which have nevertheless traditionally been involved in biotechnology, such as Japan, Israel or Canada." Or again, on the applications of *in vitro* recombination: "Many projects are under way, some of them well advanced. It must however be acknowledged that, as yet, hardly any dynamism is apparent in the field of applications."

It did indeed take longer than expected to attain the first industrial objectives, but biotechnologies have come a long way since we wrote those remarks. Moreover, it is gratifying for its authors to note that, even today, *Life Sciences and Society* contains an analysis of the development of concepts in biology and their foreseeable impact which has, essentially, lost none of its topicality.

However that may be, I recall being invited to lunch by Jack Lang, then adviser to the French presidential candidate François Mitterrand. The luncheon was also attended by Jacques Fauvet, editor of *Le Monde*, and the actor Michel Piccoli. This was in 1980, and the conversation focused primarily on the Soviet invasion of Afghanistan. Over dessert, however, François Mitterrand asked me about biology, then about what I thought of the socialist party's position on scientific research. I don't know whether my answer satisfied him, but I was subsequently invited to enter the Government as adviser to two consecutive prime ministers, Pierre Mauroy and Laurent Fabius.

The curiosity aroused in Giscard d'Estaing by a journey was shared, some time later, by his political rival. The latter was to continue to show a sustained interest in the achievements of biology and medicine, and I often had the privilege of being able to talk with him about the related problems. The seriousness and insight he brought to his consideration of these matters were always evident.

If we accept the definition given in the Spinks report, published in Great Britain in 1980, biotechnologies "relate to the exploitation of biological organisms, systems or *processes* by the manufacturing or service industries." This being so, these biotechnologies have a very long history behind them.

THE FIRST GENERATIONS
OF BIOTECHNOLOGIES

To tell the truth, what could be called the first generation of biotechnologies is more by way of being a body of traditionally handed-down empirical recipes and procedures. To find the first traces of them, in fact, we need to go back to the period which followed the Stone Age. There is reason to believe, on the basis of the study of prehistoric remains, that the first human practices relating to the conservation, preparation or improvement of food (for example, the empirical use of fermentation to produce beverages or milk products) were also the first concrete manifestations of interest in biology, and to this must be added knowledge of nutritive and medicinal plants as well as animal husbandry. Very early on, then, as far back as the Neolithic Age, people were unknowingly* making use of bacteria, yeasts, enzymes and molds to prepare food and beverages, and even to produce certain textiles.

In writing these lines, I cannot help thinking of what my colleague, the great ethnologist André Leroi-Gourhan, said about the place of technology in culture, and even in civilization: "There can be no complete picture of man without an understanding of technological man.... We are

---

* With respect to this more or less "instinctive" know-how, Jacques Robin, in his recent book *Changer d'ère* [A Change of Eras], writes: "We regard this astonishingly ingenious know-how as the forerunner of the future techniques which, almost a million years later, were to be developed in the form of tools for transforming inert matter."

living in a society which is the product of technology and their technologists. Yet, these occupy a modest place in the social scale and the hierarchy of human activities. *Homo faber* still suffers somewhat at the hands of *Homo sapiens*" (Leroi-Gourhan, *Les Racines du monde* [The roots of the world]). The studies made by the late lamented Fernand Braudel (*Les Structures du quotidien* [The structures of daily life]), dealing for example with the use of bread among the populations of the Middle Ages, or the penetrating analyses by Lévi-Strauss of the cooked and the raw, illustrate the point to which technologies, even the most modest, can bear witness to our civilizations no less faithfully than works of art or literary or philosophical writings.

But is was not until the end of the 19th century that the empirical practice of biotransformation through fermentation began to be standardized, and more or less uniform products could be obtained in a reproducible way. Pasteur's work on beer and wine probably constituted the first rational attempts to apply the biological knowledge of the day to industrial or semi-industrial processes. Brewer's yeast was manufactured, and lactic acid produced for the fermentation industries, as well as ethyl alcohol for the chemical industry. The textile industry, for its part, was already benefiting from a certain amount of knowledge in the field of enzymes, particularly of amylase. This first-generation biotechnology stemmed from the research by Eduard F.W. Pflüger into what he called living proteins, which he discovered in brewer's yeast. But, as we have already noted, it was Buchner who was the first to obtain a cell extract that could be used for fermenting sugars, thus starting enzyme biochemistry on its way. Lastly, the work of Emil Fischer was to add to our knowledge about enzymes and the possibilities of their utilization.

The First World War was the occasion for an unprecedented expansion of the fermentation industries. The manufacture of munitions calls for considerable quantities of acetone. Accordingly, the biologists of the Allied nations mobilized their experts in an effort to improve the yields of acetobutyric fermentation. A pupil of Pasteur, Fernbach, isolated strains of *Clostridia*, which performs excellently in this respect. In Great Britain a young Jewish chemist, Chaim Weizmann, also concerned himself with this problem and succeeded in developing a process for the very large-scale biological manufacture of acetone.*

During the interwar years, acetone was for some time to remain a key product of biotechnologies, primarily for manufacturing rayon, an artificial textile fiber which looks like silk. The fermentation industry was to undergo rapid expansion with the production of substances such as riboflavin (or vitamin B_2), glycerol, sorbose (the raw material for the chemical synthesis of vitamin C), citric acid, etc.

---

* His work was to have unexpectedly important consequences: in recognition of the services he had rendered the Allied cause, but also in exchange for the patent he held, the British Government finally agreed to support the Zionist project (the 1917 Balfour Declaration). Appointed after the war as President of the World Zionist Organization (1920), and then of the Jewish Agency (1929), Weizmann was to become the first President of the Jewish State and founder of Rehovot University. Forty years later, the State of Israel was to be headed by another renowned biotechnologist, also a highly regarded chemist, Professor Ephraim Katzir.

Around the 1930s, however, new chemical processes appeared which made it possible to obtain most of these compounds from hydrocarbons, and at much lower cost. As a result, petrochemistry was to become dominant. Thus by the time the Second World War broke out the fermentation industry had been relegated to a secondary role.

The discovery of antibiotics, and primarily the industrial production of penicillin thanks to the work of Alexander Fleming, Ernest B. Chain and Howard W. Florey, all three winners of the Nobel Prize for Chemistry in 1945, may no doubt be regarded as the inspiration for the second generation of biotechnologies. The considerable attention attracted by the discovery itself gave new and unexpected encouragement to the champions of the fermentation industry. Apart from the production of the other major antibiotics, bioindustry, now aided by the better domestication of microbial strains, embarked upon the large-scale manufacture of a wide range of active molecules with a high value added: enzymes, steroids, vitamins. It was only when Japanese industry entered the field, however, that manufacture of amino acids also began.

This second generation of biotechnologies, which flourished from the end of the 1940s to the beginning of the 1970s, was contemporaneous with a decisive advance in fermentation techniques: the discovery of processes which made it possible to use enzymes in the solid phase. Up until then, enzymes could hardly be regarded as genuine industrial bioreagents. Because of their instability in a liquid environment, and their steric hindrance due to overcrowding with the reaction products, they were usable only in discontinuous operations. The development of "immobilized" enzymes was to lead to a new surge of continuous fermentation techniques. Among the most remarkable achievements of these processes was the large-scale production of glutamic acid by the Tanabe Seyaku Company in 1969 and above all of penicillin, using penicillin acylase (the Toyo Zojo process) in 1976. Closer to home, we have the enormous fructose syrup market in the United States. Production of these syrups is based on a combination of two immobilized enzymes: glucose amylase, which hydrolyzes the starch in corn into glucose and fructose; and glucose isomerase, which converts the glucose into fructose, sweeter in taste than glucose.

GENETIC ENGINEERING

The first publications dealing with genetic engineering techniques date from 1972. American researchers, primarily S. Cohen, P. Berg, Helling and Boyer, turned to good use the results obtained by the Swiss biologist W. Arber, who had discovered several years earlier enzymes capable of cutting the genetic material at very precise locations, almost to the exact nucleotide!

With these enzymes, known as restriction enzymes, and a few others capable of joining DNA fragments together, as well as a number of microchromosomes which had long ago been described in miniorganisms (the plasmids), a new methodology became available which allowed "cloning," in other words the purification of genes or copies of them by inducing them to reproduce in very large numbers within host bacteria. It very soon became apparent that this recombinant DNA technique would simultaneously permit major theoretical advances and have considerable practical consequences.

On the one hand, it now became possible to use pure genetic "probes." These probes consist of amplified genes, fragments, or copies of genes (cDNA). They can be used to test, by means of a molecular molding process known as hybridization, the integrity or functionality of the corresponding gene within a cell. The probe used is most often labelled with a radioactive tracer or fluorescent agent, so that the hybrid form can be detected. This procedure was to form the basis for a large number of operations designed to diagnose hereditary diseases or infective agents.

On the other hand, genetic engineering can be used to induce a foreign gene to function within a bacterium or a eukaryotic cell. If a suitable vehicle (or vector) is chosen to propagate the gene, synthesis of a new protein can be observed in the recipient cell, which then serves as a miniature factory.

Moreover, the chemical, pharmaceutical and agro industries were to begin deriving considerable benefit from these new means of genetic reprogramming, using them to produce a large number of biological substances which could be obtained only with difficulty using conventional chemical techniques, or whose extraction from tissues was a delicate operation.

A third application of genetic engineering lies in the techniques of transgenosis, which involve the transfer of a gene into a regenerating plant cell or a previously fertilized animal ovum. In the latter case, the operation is followed by implantation in the uterus of a carrier female. As a result of the integration of the foreign gene into the inheritance chromosomes, it becomes possible to modify some of an individual's characteristics in the direction desired by the horticulturalist, farmer or zootechnician. The development of recombinant DNA techniques may thus be said to underlie the most recent generation of biotechnologies, the third so far. It also stems, however, from another cell methodology which was developed at the same time and which makes it possible to produce, also by cloning, ultrapure antibodies, known by the specialists as "monoclonal antibodies." We owe this remarkable result to the work of an immunologist of Austrian origin, César Milstein, working together with a young German biologist, Georg Kohler. Their discovery, which was to earn them the Nobel Prize, made it possible, like genetic engineering, to develop extraordinarily high-precision tools for exploring the main components of cells. It, too, led to the production of biological probes of a new kind, immunological probes this time, which are remarkably useful for all forms of diagnosis.

SOCIOECONOMIC ASPECTS

What was the socioeconomic climate in which the third generation of biotechnologies developed? What were their first concrete spin-offs? Where do we stand today? The answers to these questions will lead us to discover a completely new concept of biology, a concept which may be termed commercial. The business world is extremely interested in the development of these lines of research, and biologists obviously have to take this interest into account.

The reactions of industrialists, investors and bankers to the first successes of genetic engineering, despite the fact that they were essentially theoretical in nature, were to take the form of an unprecedented infatuation. Between 1976 and 1987, for example, a rapid increase took place, especially in the United States, in the number of companies focusing specifically on the applications of this new research. Today there are almost two hundred of them, though indeed of widely varying size. Within an extremely brief space of time, a number of these companies came onto the stock market, and the price of shares climbed. The immediate result of this was an overvaluation of the real economic capacity of most of the firms newly established between 1977 and 1981; in the latter year alone forty-three companies emerged.

Around 1983, therefore, a certain amount of disenchantment set in. Expectations regarding the speed with which results would be achieved had probably been too high, with the public counting on the appearance of new miracle pharmaceuticals. It was the time when people were singing the praises of interferon, a protein of cellular origin that inhibits intracellular multiplication of a large number of viruses, which it was thought would be of use in the treatment of cancer.

Not only did the companies born of this first burst of enthusiasm not make any profits, but many of them went under or were taken over by larger companies active in traditional fields (chemistry, engineering, the agro industry). There thus appeared to be grounds at that time for fearing that the business world had been seduced by the siren song and that the supporters of biotechnology had merely been engaging in wishful thinking. For this reason, the years 1984–1985 saw a sharp downturn in financial investments.

And yet it is apparent today that, while the enthusiasm shown at the time was excessive, the movement thus initiated—a kind of *Sturm und Drang* of advanced technologies—proved extremely positive in the medium term. It at least paved the way for the economic recovery and gave the United States an undeniable advantage over its competitors.

Jean-Jacques Salomon, in his book *Le gaulois, le cowboy et le samurai* [The Gaul, the cowboy and the samurai] (CPE, 1985) admirably expresses the conclusions that can be drawn from the experience of the first joint-venture companies: "Even though it is true that speculation has already experienced ups and downs, with some of these firms going bankrupt or being bought out by the big corporations, it is thanks to their spirit of enterprise that research and marketing efforts are most dynamic in the United States. Indeed, it appears that there is a kind of division of labor, with the small private firms specializing in the applied-research phase of product and process development and the large corporations taking charge of production, marketing and regulatory negotiations."

But the American adventure is of interest in still another way: it gave rise to the creation of a new type of structure, midway between the university and the industrial world: the start-up companies. Between 1977 and 1984, initiatives were frequently taken of a kind which until then had been extremely rare, with leading researchers, for the most part very well known in their disciplines for basic research, deciding to become the main motive force behind companies or even to set up their own. Examples are Walter Gilbert and David Baltimore in the United States, both of them Nobel Prize winners, and Charles Weismann, an eminent molecular biologist working at the Zürich Polytechnicum.

Following their lead, the research world showed such enthusiasm for developing new companies and for transfer activities that this movement ultimately gave rise to some concern on the part of United States Federal authorities (the National Institutes of Health, the National Science Foundation and the National Academy of Science). Was there not a risk, if research took this turn, that free scientific communication would be endangered, with some biologists preferring to refrain from publishing their findings so as to retain priority for commercial purposes?

It should be noted, however, that in general the dynamics of basic research apparently did not suffer on account of this state of affairs. American researchers, unlike their European colleagues, are blessed with a special procedure whereby their work does not fall into the public domain until a full year has elapsed since the date on which it was published in a scientific journal. This is known as the grace period.

The disenchantment of the mid-1980s proved to be of short duration. A balance was speedily restored in research and development investment procedures. Fearing that some of their products might become obsolete or that they might lose some of their markets, large multinationals which had hitherto stood aside from this sector took the plunge on a more or less massive scale. Japan and Europe, after a relative wait-and-see phase, soon followed suit, to the point where most experts currently agree that a new era has begun, the era of consolidation.

This diagnosis is supported by a whole series of factors. Most significant among them, no doubt, are the arrival on the market of the first major products spawned by the new technologies, and the forthcoming emergence of a whole range of substances which have already been tested at the clinical stage or proven their agricultural advantages. These substances are likely to be granted marketing authorization in the near future.

Among the grounds for hope, we must also mention what could be called the awakening of certain developing countries to the applications of biotechnology. For example, large numbers of bioindustrial activities are being started up, often backed by local or bilateral intercompany agreements and corresponding to well-defined national programs. Lastly, this statement of reasons must be rounded out by reference to the extremely rapid progress or refinement of biological research techniques, which certainly herald the advent of a fourth-generation biotechnology involving protein engineering.

What, then, are the new products? If in 1982 the American firm Eli Lilly and Company obtained, without much difficulty, authorization from the Food and Drug Administration (FDA) to market insulin produced by genetic engineering, this was surely not only because the need for insulin to treat diabetics was increasing worldwide but also because the hormone's therapeutic properties were already well known. The other products could not be marketed as quickly. The rise of the environmental movement no doubt played a part in this. In any event, it was not until the end of 1986 that the first significant range of substances produced by means of this new technology became available.

Among the main products commercially manufactured for use in the pharmaceutical sector, some, such as insulin, the new vaccines against hepatitis B and the human growth hormones, were replacements for existing products hitherto manufactured by more conventional processes; others, on the other hand, such as the specific tissue plasminogen activator (TPA), a powerful antithrombotic, alpha interferon or the monoclonal antibody designed to slow rejection reactions to grafts (Ortoclone), were not yet being produced industrially. A look at the sales figures for each of these products indicates that the huge financial markets the experts had predicted in the early phases of biotechnology are still far off. Nevertheless, these figures are by no means negligible, since they now total in excess of $500 million, whereas in 1985 total gross sales of biopharmaceuticals derived from genetic engineering amounted to only $65 million. It must also be taken into account that some of these new bioproducts have been manufactured on an industrial scale for barely two or three years.

Moreover, there is another list that needs to be taken into consideration, that of the substances for pharmaco-clinical use which have already met the main requirements of therapeutic tests and will shortly be marketed following the necessary regulatory approval, probably within one or two years, some of them perhaps sooner. For the most part, these comprise a family of molecules belonging to the category of immunomodulators. These compounds, which act at the lymphocyte level and are generally known as inter-leukins, behave as agents capable of activating the immune defenses. The list also includes growth factors such as the compound EGF (epidermal growth factor), which should lead to innovations in the treatment of severe burns.

It is thus apparent that, contrary to what is often believed, biotechnologies not only create substitution products, but are also well on their way to developing molecular tools hitherto unavailable to medicine. A count of all the new products derived from biotechnologies and in the process of being marketed, in the United States and Europe as well as in Japan, yields a total of no less than two hundred proprietary products (corresponding to fifty different molecular entities) which will be available within the next three years.

Certainly, if there is a field in which forecasts of all kinds are rife, this is it. The history of the past decade counsels the most extreme caution where the quantitative side of estimates is concerned. But the margins of error are narrowing, now that the first orders of magnitude of the markets for the key products and the list of derivatives about to be marketed are known.

According to a recent estimate by an American firm, Boston Biomedical Consultants (1986), the world market for products created by the new biotechnologies in the public health sector alone came close to $4.2 billion by 1990; the bulk of this total would comprise products for pharmaceutical use ($2.9 billion), the remainder stemming from the sale of monoclonal antibodies. Other consulting firms, such as CRC (Consulting Resources Corporation), are somewhat less optimistic, putting forward a total figure of $1.7 billion for the same time period.

All the evidence indicates that the pharmaceutical industries are entering a new phase and that biotechnologies are going to earn them fairly substantial profits. As for longer-term forecasts, it is estimated, again for the pharmaceutical sector (therapeutic and diagnostic agents as well as vaccines), that the world market should exceed $15 billion by 1995 and approach $50 billion by the end of this millennium.

But biotechnologies are multisectoral, and their applications are likely to become increasingly important in areas other than health. Since at present the market for bio-industries deriving from new technologies aimed at agriculture or the agro-industry remains small (as marketing licenses have not yet been granted), all we have to go by are expert estimates. According to SRI (Stanford

Research Institute, 1987), by the year 2000 agribusiness products developed by biotechnologies could run up sales of $9 billion, or 30% of the total market. Biotechnological agriculture, on the other hand, with a market of $5 billion, would cover only 3.3% of the sector. The equipment and instrumentation fields are expected to reach $4.8 billion, and pollution control and chemistry, $2 billion and $0.3 billion, respectively.

GREEN BIOTECHNOLOGY

While technological transformations are now in full swing in the biopharmaceutical field, the foundations are being laid for far-reaching changes in plant biology, changes which are regarded in some quarters as more important and likely to have a greater impact than those relating to the study of microorganisms or animal cells. Will this, perhaps, prove to be the true technological revolution of the next millennium?

The plant kingdom has long been the poor relation in the modern odyssey of biology. The bacterium was, as we have seen, the *ens rationis*, the biological factory, on which researchers were able—as indeed they still are—to try everything, or almost everything, but it does not truly embody the *natura naturans*. Since it reproduces identically in an almost infinite manner when the nutritional conditions are suitable, it became the ideal model for the adherents of biometry and molecular biology. The animal cell, as soon as it was able to meet the requirements of *in vitro* culture, was able in some cases to supplant microbes, since there was indeed a need to probe the workings of the eukaryote world. The plant cell, for its part, remained in the background, for a number of reasons relating to the properties of plants: their very luxuriance, their polymorphism, the complexity of their genomes and the difficulties attached, at least in the beginning, to the study of cell populations; to which must be added the existence of a cellulose wall which makes them impermeable to most chemical reagents. No doubt the initial attitude of botanists towards the advances in molecular biology also had something to do with it.

Is this attitude to be attributed to the singular nature of the subject of their study, or more prosaically to the fact that they had long ruled the roost in the academic world, and had every intention of continuing to do so? Or was there a more obscure reason: the implicit desire to protect that prime symbol of living nature, the plant world, so dear to comparative naturalists? The fact remains that they showed little alacrity in accepting the ideas and above all the methods of molecular genetics. That time is now, I think, fortunately past.

Without wishing to retrace the history of the advances in our knowledge of plant physiology, let us recall that back in the mid-1950s the French botanist Roger Gautheret succeeded in developing the first cell cultures in carrots. Jean-Paul Nitsche demonstrated a few years later the remarkable regenerative power of the plant kingdom: under appropriate conditions, an embryonic somatic tissue, a meristem, or even gametic cells can reconstitute

an entire plant! At the same time, the possibility was discovered of converting a plant cell into an element with no
walls known as a protoplast, a remarkable artifact capable
not only of regenerating its wall, but also of merging with
a protoplast taken from a cell of a different species. This
somatic fusion thus makes it possible to break down the
barriers against interspecies crosses.

From the beginnings of genetic engineering, many
researchers dreamt of transposing its methods to the manipulation of plant cells, but they were confronted by a serious
problem: appropriate vectors capable of introducing a foreign gene into the plant cell were not available then. This
obstacle was removed in the 1980s, when the Belgian
school headed by J. Schell and M. Van Montagu showed
that plasmids, those circular chromosomes present in the
bacterium responsible for crown gall disease in tobacco,
were remarkable gene vectors once tumorigenic DNA segments had been removed from them by enzymatic surgery.
The possibility was thus afforded of transfering new genes
in monocotyledons.

The case of dicotyledons proved more difficult, since
the plasmids of the crown gall bacterium, *Agrobacterium
tumefaciens*, cannot integrate into their genetic material.
Indirect approaches thus had to be resorted to: encasing the
genes in liposomes (artificial sacs having the ability to
penetrate into cells), forced penetration induced by electric
shock (electroporation), etc.

Today, transgenosis can be performed on plants using
as the receiving cell a simple suspension of somatic cells
taken from the plantlet in its embryonic state. Since these

cells are capable of regenerating an entire plant, this artificial transgenosis makes it possible to increase the natural variability of the plant kingdom. Lastly, the cloning in plant (or other) cells of various plant genes is beginning to shed valuable light on the delicate organization of their chromosomes, including their regulatory regions.

As progress is made at the basic level in the cellular and molecular biology of plants, the first major applications are found to be emerging at the biotechnological level. One of the first objectives involves, for example, the large-scale reproduction of plants possessing previously selected properties. This goal is often difficult to attain using the usual methods of propagation from seeds, because of the many variations that can occur in the plant in the course of its development. This is why, for about a decade now, the clonal propagation technique has often been employed. This micropropagation makes it possible to obtain rigorously identical individuals. In the beginning, the technique was used primarily to obtain homogeneous populations free of viruses, since the individuals that reproduce in this manner are generally resistant. Subsequently, its application was extended to the cultivation of plants without the use of seed. Clonal propagation, which is particularly well suited to horticulture, is also used for edible plants, such as potatoes, as well as in forestry. Spectacular results have been achieved on oil palm plantations. Better still: in an effort to use the regenerative capacity of plants to best advantage, biotechnologists have come up with the idea of gradually replacing the use of natural seeds by artificial seeds. This involves nothing less than growing somatic embryos under conditions strictly controlled by humans. For this purpose, undifferentiated plant cells cultivated *in vitro* are encased in an artificial envelope which protects them against pests or predators for so long as the tissue has not acquired natural resistance.

Although this process has yet to be placed on a commercial footing, extremely encouraging results have been obtained by researchers at the National Agronomic Research Institute for rice, alfalfa and cotton. In 1986, a research project was launched under the EUREKA program with joint participation by Limagrain, Rhône-Poulenc Agrochemistry and Francereco (the French development center of Nestlé). Its purpose is to assess the industrial cost of using artificial seeds on a large scale, with a view to identifying economic targets.

The aim of micropropagation techniques, as we have seen, is to perpetuate a given genome. Conversely, the most frequent objective is to modify the genome of plants, to increase their variability. This constitutes an artificial revival of the old technique of grafting, which was also designed to bring out a whole range of favorable characteristics.

The first way of doing this is to fuse the protoplasts of different species. The hybrids thus created sometimes show curious properties, as in the case of the "pomato," a hybrid between a potato and a tomato which produces tubers in the soil and fruit above it! Unfortunately, the resulting hybrid is sterile and of arguable agricultural interest. Nevertheless, this technique has had a few genuine successes: in the hands of Japanese biologists, somatic hybridization has made possible the creation of new forms of cabbages with food qualities that are highly prized in Asia and are being exploited commercially (Biohakuran, Senposai, etc.). The hybrids of radishes and cabbages produced by the French National Agronomic Research Institute also seem to show promise.

However, the high road to increasing plant variability seems to be the gene transfer technique, the principle of which we have already outlined. Although none of the plants modified in this way are yet being used in horticulture or agriculture, and although despite field tests most of the results obtained have not yet gone beyond the experimental stage, they are already revelatory of the vast potential offered by genetic engineering. Initially, an effort is made to identify the most interesting gene candidates for such transfer tests. Already characterized as such have been the majority of the genes known as structural genes which code for the production of the major nutrient proteins, for example phaseolin in peas, leghemoglobin in legumes, etc., have already been characterized as such candidates.

In addition, while data are still lacking with respect to the regulator genes, substantial progress has nevertheless been made: those which play a key role in floral differentiation, the formation of receptors for auxins (growth-promoting substances in plants) and the control of membrane fluidity have already been identified. Regulatory sequences have also been discovered which are capable of stimulating the action of structural genes from a distance. Since these elements of the genome, which have been given the name enhancers, are capable of stepping up the activity of specific genes within given tissues or specific organs (leaves, roots, etc.), means will shortly become available of inducing targeted activity of a gene introduced into a clearly defined region of the plant.

A good number of the projects developed in industrial companies and drawing on transgenosis are well advanced. An example is the artificial transfer of genes capable of neutralizing the functioning of polygalacturonase, an enzyme which dissolves pectins (substances responsible for hydration of plant cells) in fruits; tomatoes with a long shelf life have been produced by this means. The main effort, however, is directed at improving the high-nutrient-value protein content of plants (proteins with a high sulfur or lysine content). Other projects relate to horticulture: for example, transgenosis has already produced petunia varieties of hitherto unknown coloring.

But the interest of the new genetics in this field does not stop there. It also gives an insight into the major problems of plant physiology: nitrogen fixation, mineral uptake from soils, or protection against plant pests.

Where nitrogen is concerned, the importance of the use of chemical fertilizers for the food economy is well known. Their manufacture currently accounts for more than 30% of agriculture's energy bill. Ten years or so ago, French consumption already amounted to two million tons. But nitrogen fixation systems occur in nature. They are found in the *Cyanobacteria* of rice paddies, in photosynthetic bacteria (photosynthesis being, of course, the group of operations which uses solar energy to make possible the manufacture of organic matter), and above all in the genus *Rhizobium*, which includes microorganisms that live in symbiosis with the host plant. For example, *Rhizobium*-legume symbiosis accounts for half of nitrogen fixation planet-wide (clover, soy, alfalfa, peanuts, beans, peas, etc.). As we know, bacteria frequently take as their biotope, that is, as their living environment, the rootlets of these plants where they develop nodosities and thus operate in a semianaerobic state. (In the absence of oxygen from the air, some microorganisms derive oxygen from inorganic chemical molecules and use it as a hydrogen acceptor: this is what is known as "anaerobic" respiration.) They transform nitrogen from the air into nitrate through the use of a large number of enzymes and carriers.

Most of the genes responsible for nitrogen fixation have now been successfully cloned. Even though this highly complex genetic system is far from being fully understood, the economic and social consequences that would result from complete mastery of the mechanisms involved in fixation of the Earth's nitrogen—and, if I may say so, the posthumous joy of Justus Liebig and Jean-Baptiste Boussingault, who 150 years ago pioneered the study of the nitrogen cycle—can be readily imagined.

It is no doubt in the field of acquired resistance to pests that the most rapid progress can be expected, leading to the first plants artificially modified by genetic engineering techniques. American enterprises have in fact devoted considerable effort to confering new herbicide-resistant properties on horticultural or agricultural varieties. One of the strategies for attaining this objective involves transfering to the plant a bacterial gene responsible for resistance to the chemical in question. Thus, the Monsanto Company has succeeded in producing petunias that are resistant to one of the most commonly used herbicides, glyphosphate, through the transfer and incorporation of the EPSP gene; Calgene has achieved the same success with poplars. For its part, Dupont de Nemours, a giant of the chemical industry (which in 1984 was already spending more than $100 million on biotechnological research and development), has succeeded in producing plants that are resistant to sulfonylurea by transfering to them the bacterial gene that codes for acetolactate synthase.

As for phytophage, natural agents which attack plants and cause their diseases, or even their virtual disappearance (one need only think of Dutch elm disease), these are not in short supply, be they viruses, bacteria or molds, or helminths, insects and birds. For a number of years now, every effort has been made to understand the intimate mechanisms of this relentless biological struggle in order to turn its weapons to use in plant protection. While pyrethroids and other plant pest control agents of chemical origin continue to be of immense service, many efforts are under way to enlist pheromones, the hormones which enable insects, *inter alia*, to communicate at a distance. One of these efforts involves the manufacture of chemical lures to divert harmful insects.

The release of sterile males is another strategy, nor should we forget the traditional use of ladybugs to control aphids. Modern biotechnologies will certainly have a role to play in protection against phytophages. The best illustration of this today is the transfer to tobacco plants of a gene from *Bacillus subtilis*. In its normal host, this gene is responsible for the synthesis of a bizarre protein which has been extensively studied at the Pasteur Institute (R. Dedonder, de Barjac), known as the crystal protein because it can crystallize right inside the bacterium. The molecule of this crystal is one of the most powerful insecticides known, and also one of the least harmful to humans and other mammals. Tobacco plants which have artificially incorporated the crystal gene are naturally resistant to most phytophagous insects, and the effect is hereditary.

This has no doubt been a rather long treatment of the plant question; but it will be apparent that I more or less share the opinion expressed by François Dagognet in *La maîtrise du vivant* [The mastery of life] (Hachette, 1988) that the plant kingdom, hitherto neglected by molecular biology and relegated in an excess of enthusiasm to insignificance, is undergoing a rehabilitation. There can be no doubt that plant life is called upon to occupy the leading place "both in the biological factory and in tomorrow's society."

BIOTECHNOLOGIES AND GEOPOLITICS

But it is not only this formidable "green" biotechnology that justifies our interest in this aspect of biotechnologies. Their development may have considerable geoeconomic and geopolitical consequences. We must not be afraid to acknowledge that the face of the Earth may be changed as a result.

For a long time, it may have been thought that this was only the concern of the major powers: a deft parry, as it were, of environmentalist pressure and at the same time an economic strategy for standardizing the production of molecules with a high value added at the expense of the developing countries. There were also grounds for fearing that, monopolized as they were by the three major industrialized blocs, biotechnologies might be turned to the task of accelerating the downfall of the Third World countries. Albert Sasson has explained this very well in *Le jaillissement des biotechnologies* [The flood of biotechnologies] (Fayard-Fondation Diderot, 1987) by reference to the techniques of micropropagation and manufacture of molecules of pharmacological or nutritional interest from cultured plant cells.

Left to the multinationals alone, which moreover maintain collections of strains, seeds and genes, these techniques would lead to the replacement of natural resources coming from the southern hemisphere, and hence of the products derived from them in the northern hemisphere, by "standardized" molecules whose source would be independent of all imports.

The risk remains, and it should not be underestimated. Conversely, however, a future which favors the Third World countries may emerge if they make a firm commitment to the dynamics of biotechnologies, while making choices compatible with their economies and their local resources.

This is a problem to which, not too long ago, Hubert Curien, Philippe Kourilsky, Jean-Claude Pecker and I drew President Mitterrand's attention: some biotechnical solutions already exist; others may be within reach; advanced technologies of different kinds (for example the use of observation satellites, geochemistry, or hydrology) could extend their field of application by assisting in the better definition of arid zones or predicting the dynamics of plant or animal populations. But the limiting factors are still many: first, there is the problem of the critical mass of "brains" (researchers, engineers, trainers); more important, there are a series of questions relating to geopolitical and geoeconomic balances which cannot be analyzed in detail here. The major powers need to understand that "impounding" advanced technologies in order to keep the benefits to themselves cannot fail to destabilize the system of trade in goods and services. They will then see a substantial part of their markets dry up, not to mention the serious debt problem which will continue to delay any change in the economic dynamics of the southern hemisphere.

François Mitterrand seemed quite sensitive to these arguments, and indeed partially echoed them in his *Letter to the French People.*

It does appear that we are, happily, witnessing a measure of biotechnological and bioindustrial revival within many countries located outside the bloc of the three major economic poles. Admittedly, as the figures clearly show, the latter still retain almost 90% of world biotechnology capacity. But the will to redress the balance has been perfectly apparent in the rest of the world for the past two or three years, as can be seen in Asia, but also in Latin America.

The case of China is particularly striking: the world's most populous country needs to tackle a multitude of problems in order to ensure its supplies of energy, food resources and medications. Under the auspices of the State National Committee, very extensive efforts are being made in *in vitro* plant cultivation, selection of hog breeds, production of biogas from liquid manure (digesters), fermentation processes and enzyme production. In 1986, moreover, the Government of the People's Republic of China set up three national centers, at Shanghai, Beijing and Canton, designed to speed up the development of biotechnologies. The still more recent opening of the Shanghai Biological Processing Center will give this great Chinese community a vast arsenal of techniques focusing on bioconversion processes. Aware of the need to step up the tempo, China is setting up various bioindustrial companies, in most cases with substantial foreign investment. In three years, the Chinese have thus quintupled the production of enzymes for use in science, the agro-food industry and medicine. They also aim to achieve a major degree of self-sufficiency in vaccine production, and are already working very actively on the development of an anti-hepatitis B vaccine produced by genetic recombination.

In Latin America, although the continent has not succeeded in establishing the equivalent of a Common Market to exploit its natural resources, the private sector has gradually come to understand the need for a policy of cooperation and integration in order to make up for the precarious supply of essential investment capital. A fairly large number of transnational consortiums are thus being created, among them the huge conglomerate of the Andean Pact Financial Corporation, recently established to exploit the cloning of higher plants and develop shrimp aquaculture using modern techniques; or, again, a major bilateral structure involving Argentina and Brazil, the Argentine-Brazilian Biotechnology Center (CABBIO), which is working on projects of common interest in areas relating to the production of new forms of energy, health, animal husbandry and the agro-industry. In addition, half a dozen major bioindustrial enterprises have seen the light of day in various Latin American countries, frequently affiliated with North American, Japanese or European interests.

Many more examples could be given. Countries such as India, Korea, Taiwan and Singapore are investing a great deal in biotechnology. In Africa, although a slower start has been made, several national programs are beginning to yield concrete results, notably in Tunisia.

Before abandoning these macroeconomic analyses in order to go on to discussions of a completely different order, it is probably helpful to give a brief indication of the current balance of forces within the group of industrialized countries.

Taking into account both the number of industrial companies and the level of public and private investment for the period 1983–1986, the United States may be said to lead the field, with 40% of world potential, followed by Europe (30%) and Japan (20%).

Within the European Economic Community, if the situation is viewed in terms of public expenditure, France places third, behind Germany and the United Kingdom.

It must be added that some sectors of the world market are dominated by certain national companies. Thus France is particularly far ahead in the plant seed sector, where it is nevertheless losing ground to Switzerland and the United States. Great Britain excels in the field of monoclonal antibodies, and has established bioindustries on a risk-capital basis. Germany, while suffering from a relative shortage of experts, has concluded agreements, which have been variously received, with the giant Hoechst International, as well as with a number of American universities, to develop the agro-food sector and second-generation drugs. Denmark, with its Novo Company, and the Netherlands with Gist Brocade, are today the world leaders in enzyme production.

Further impetus will come from the initiatives taken by the authorities of the Community itself, particularly by the Bridge Program (Biotechnology Research Program for Innovation, Development and Growth in Europe), which started up in 1990 and already has financing to the tune of $120 million.

Lastly, let me dwell for a moment on the case of France, inasmuch as a number of general conclusions can be drawn from it with regard to research policy. Mention should be made of the report commissioned by Valéry Giscard d'Estaing, followed by the technical update by Joël de Rosnay, and then by the lengthy study by Jean-Claude Pelissolo. While timely attention had been drawn to the importance and urgency of the problem, it was not until 1982 that real political will resulted in a concrete follow-up of the experts' work. The development of biotechnologies is among the seven mobilizing programs launched by Jean-Pierre Chevènement.

An impatient decision-maker, the Minister of Research and Technology did not hesitate to prod the Government a little in order to speed things up where biotechnologies were concerned. Needless to say, he had my support as adviser to the Prime Minister. Both Salomon in his book and Jean Dausset in a report drawn up in 1985 noted, three years later, that this program had encouraged numerous positive actions.

It is easy to view this centralized and programmed mode of State intervention with irony, rightly arguing that private initiative should be allowed the greatest freedom and spontaneity. But in a sector like biotechnology, which had not yet—far from it—reached its maximum expansion rate and which, moreover, was the battleground of fierce international competition, such public aid was needed, and it did indeed prove effective. It was thus a mistake on the part of Jacques Chirac's Government to cut the budget package for biotechnological research so steeply, from 120 million to 30 million francs. Made in the name of some kind of ultraliberal dogmatism, it also led the Government, in the spring of 1986, to cut the overall research budget by 3 billion francs.

Even today, the consequences of this blunder are still difficult to assess, but they will be severe. And they would have been even more severe for biotechnology were it not for the determination of scientists like Daniel Thomas and Pierre Douzou, who was responsible for the mobilizing program. The study entrusted *in extremis* to René Sautier, former President of Sanofi, by the Prime Minister at the end of his mandate may undoubtedly be taken as belated recognition and a sign of definite concern for the consequences of a decision which, to say the least, was too hasty.

Many other aspects of the development of biotechnologies would no doubt have been worth describing and discussing here, such as legal and ethical aspects, ideological aspects. Without exhausting the field in this book, we shall have occasion to return to them individually from other angles.

III

HEREDITARY

DISEASES

AND ETHICAL

ISSUES

If there is one field in which biology and society should engage in a dialogue, it is, supremely, that of human health. Nothing new in that, it may be said. On the contrary, I should like to show that the terms in which the problem has to be posed have radically changed. In fact, the time that separates a discovery in fundamental biology from its application to public health is becoming increasingly shorter. The public and officials are now peering over the physician's shoulder towards the laboratories where basic research into living matter is going on.

This is understandable: within the space of a very few years, molecular biology, while continuing the fundamental study of living beings, but this time in the "higher" organisms so called, has upset most of the classic concepts of pathology and even in some cases the professional approach taken by physicians.

These remarks apply above all to the recent technical breakthroughs in molecular genetics and immunology, which are resulting in diagnostic methods of a precision unimaginable only six or seven years ago; but they also apply to the advances made in the biology of human development and reproduction.

We owe the term "predictive medicine" to Jacques Ruffié, a leading specialist in human typology, who is interested in the genetic markers carried by blood groups and other blood cell antigens in particular. It has become still more prominent following Jean Dausset's discoveries in the field of histocompatibility antigens, sometimes known as grafting antigens, and above all since genome analysis, through its restriction mapping, has made an inroad into the precise forecasting of disease risks. The control of human reproduction, to use a term dear to Professor Jean Bernard, represents, together with the power of curing and the power of knowledge, what—at least for the sphere of biology and medicine—he terms the "third power."

And so biology, whether it deals with the gene or the neuron—henceforth its two preferred subjects—has become the business of society. On the one hand, both the healthy and the sick are inclined today to look to biology, from which they expect, no doubt too optimistically and too soon, an explanation of their individuality and their destiny. But in addition, and this is no less remarkable a feature of our civilization, biologists now maintain a relationship of constant dialogue with a society to which they in turn, in a sense, are constantly looking to for answers. What may be termed the implicit social contract, which in the 19th century and the early years of the 20th governed the relationships between biologists and the surrounding world, was aimed at isolating the former, whether through deliberate action or by tacit consent. The only good scientist was one shut up in an ivory tower, enclosed within what Pasteur called a "temple of the future," deaf at least while doing experiments to the echoes of tumult which sometimes penetrated the ears from outside.

Nowadays, a different kind of contract, also of course a tacit one, seems to have been concluded. The public at large has rarely been seen so passionately interested in the life sciences. But still more rarely have so many biologists and physicians been seen taking part as full-fledged participants in the social debate.

A curious process of substitution is even apparent: lawyers, for example, are standing by almost passively as progress in genetic engineering shatters the rules relating to patents, and philosophers for the most part are doing no more than listening in silence to the views expressed by biologists on the nature of life. There is nothing, not even the debate on ethics, that is not in large measure being called for, if not conducted, by biologists themselves.

Some may think that the attitude of biologists is simply a means of letting off steam: long regarded as genteel dreamers, they are no doubt not unhappy to come downstage a bit and cross a few swords with their colleagues in the other "exact" sciences.

In my view, however, a deeper explanation is needed for this sudden and growing propensity for contact with the outside world. This is what Jacob, Royer and I suggested at the end of our report to the President of France some ten years ago: "Contrary to what people would have us believe, it is not from biology that a particular idea of humankind can be derived. Rather, it is on the basis of a particular idea of humankind that biology can be enlisted in its service."

We are still too strongly haunted by the spectre of the last world war. Biologists, geneticists and neuropharmacologists have not forgotten the terrible Nazi experience. For a biologist to ponder the social impact of the new prenatal diagnostic techniques, genetic imprinting or medically assisted reproduction is thus not simply a flirtation with moralizing or Manicheism.

We all know today, our fears made stronger by history (and also by what we see around us), that it takes very little to tip the balance of medical practice, born of the progress in biology, over into irrationality or pure and simple ideological excess. And even if we consider these extremes to be, fortunately, neither on the agenda nor the immediate horizon, we still have every right to ask ourselves what we should do, now that we have been given this "new power" of which Jean Bernard speaks, and to wonder about the type of society we want to help build. It is with these questions in mind that we shall now turn to some of the issues which, for the past few years, have been held to form part of bioethics.

MOLECULAR BIOLOGY AND HEREDITARY DISEASES

If I have chosen to approach these issues by way of the contribution molecular biology has made to the treatment of hereditary diseases, it is not out of a desire to shock, but to encourage the reader to take a balanced view.

Too often, where these issues are concerned, analysis focuses and then fixes exclusively on the emotionally supercharged case of artificial procreation. This results in extremist general positions and often rash speculations, whether on the side of moral and social conformism or of scientific utopianism. I propose to keep intact the thread linking the undeniable and striking medical progress that has been apparent for the past few years with these ethical issues.

Any view or argument which would have the effect of slowing down this progress, impeding it or diverting it from its course, must, it seems to me, be rejected. This, in any event, is the criterion of judgment I would suggest we apply, after I have shown how this progress not only obeys a kind of Promethean logic which seeks to compel nature to yield up its secrets at any cost, but also opens up radically new possibilities of easing or preventing atrocious suffering which just a short while ago was irremediable or inevitable.

I shall not recapitulate here, even in brief, the history of the concept of disease; suffice it to recall that it was only at the beginning of the 20th century, following Pasteur's discoveries, that the concept of "diathesis" emerged and that people began to speak of the "genetic factors" in certain diseases. More precisely, it was in 1901, in the course of analyzing the metabolism of two amino acids, phenylalaline and tyrosine, in individuals showing evidence of serious hereditary disorders, that Garrod put forward the hypothesis of a direct relationship between genes and enzymes. For a long time, research into hereditary diseases was to involve essentially observations in teratology, that is to say the acknowledging of congenital malformations, or at best full-scale studies of the transmissibility of genetic taints, as they were then called, such as the celebrated hemophilia syndrome in the Russian royal family.

The modern approach to the problem was to coincide with the refinement of cytological and karyotypic studies, but above all with the work of the American school led by Linus Pauling and Vernon Ingram on "molecular diseases" (1957). It was then demonstrated for the first time that a well-defined hereditary disease affecting the red blood cells, namely sickle-cell anemia, was due to a point mutation. This simple change of a single "link" in the protein resulting from the mutation leads to the clumping of the entire hemoglobin complex, ultimately resulting in a sometimes fatal obstruction of the blood vessels. Subsequently, human genetics, by localizing on the chromosome map the mutations that give rise to many hereditary diseases, and then using recombinant DNA techniques, brought about decisive progress in this particular field of pathology.

Hereditary diseases, although relatively rare, are nonetheless fearsome. In his book *Environnement et médecine* [Environment and Medicine], Dr. Maurice Stupfel writes these lines on the subject: "In current medical treatises, these include malformations of the skeleton (brachydactyly, polydactyly, phocomelia, harelip, dislocation of the hip, dental malformations) and disorders of the skin (Recklinghausen's disease, ichthyosis, palmoplantar keratoderma), of the muscles (myopathy), of the sensory organs (color blindness, blindness, deafness) and of blood coagulation (hemophilia). The advances in biochemistry and immunology have extended this list almost infinitely, because they have brought to light the possibilities of hereditarily transmissible conditions that are less apparent since they are determined by enzymatic or immunological malfunctions (glycogenesis, cystinuria, hemo-globinopathy and conditions associated with the HLA factor)." An impressive enumeration, indeed, of this terrible genetic burden. But to this already impressive list should be added the many disasters of hereditary origin that can arise during the embryonic development of the nervous system (anencephaly; non-closure of the neural tube, or spina bifida; iniencephaly or the "no neck" syndrome, described by Klippel and Feil in 1912, etc.); not to mention manic-depressive psychosis, Huntington's disease, Alzheimer's disease, the many causes of changes in the neuronal or glial metabolism of the glucocerebrosides (Gaucher's disease) or the gangliosides (Tay-Sachs disease), Wilson's disease, Lesch-Nyhan syndrome, which is sex-linked, trisomy 21, etc.

Enough of this painful litany. It will come as no surprise to learn that almost 3,000 diseases linked to genetic defects are now known. In all, 1% of births bear their signature, and almost 30% of infant mortality (in the countries not affected by major infectious endemics or malnutrition) is due to diseases of this kind.

As has been said, the location of the gene in question on the chromosome map has been identified for only a few dozen of these diseases. In some cases, we are able to specify the chromosome on which the mutation is found, and even the chromosome region, provided the target gene has been identified. This molecular genetic process, known as reverse genetics, in which the biologist in a sense proceeds from the disease to the gene and then from the gene to its function (rather than from knowledge of the protein modified to identification of the gene), has made it possible, among other things, to establish that the "morbid locus" of as severe a disease as mucoviscidosis, or cystic fibrosis of the pancreas—in which affected children choke to death through overproduction of mucus—is carried by a specific region of chromosome 7. In other cases, not only has the gene carrying the locus in question been identified, but it has even been possible to isolate this gene by means of genetic engineering techniques. This in turn has made it possible to identify the corresponding protein and, thanks to the production of specific antibodies, to locate the modified protein within the cell and even detect its functioning.

This striking success has been achieved in the study of a very severe hereditary disease which affects the striated muscles, that is, the muscles in which contraction is voluntary and the heart muscle. This condition is known as Duchenne's disease, from the name of the French doctor (Duchenne de Boulogne) who identified it in 1866, or Duchenne's muscular dystrophy (DMD). Its clinical picture is characterized by the onset of difficulty in walking at about the age of 3, and by progressive muscular atrophy which condemns children to the wheelchair at the age of about 10 and then to respiratory disorders in adolescence followed by death at around age 20. The work of Kunkel and his group (1986), preceded admittedly by an impressive body of clinical, genetic and biochemical data from a very large number of laboratories throughout the world, led after forty years of painstaking work to the complete isolation of the DMD gene. Incidentally, this gene is the largest of all known genes (almost 2,000 kilobases, or one third the size of the complete *Escherichia coli* chromosome, which we know to contain almost 2,000 genes). This "macrogene" has been almost completely sequenced, with the result that its gene expression product, a huge protein known as "dystrophin," has also been described in detail. Thus for the first time, within an extremely brief period (three or four years), "reverse genetics" has succeeded in elucidating the molecular and cellular mechanism of a disease.

The frequencies with which hereditary diseases, at least those that are reported to doctors, "surface" vary considerably in cases in which the disease occurs as the result of a single change (conditions known as monogenic): some, for example, are particularly rare, with a prevalence of less than 1 in 100,000, while others, such as mucoviscidosis or DMD, are much more frequent (1 in 1,600 and 1 in 3,500 respectively).

But many of these conditions are recessive: they do not manifest themselves in carriers of the mutation provided that one of the two parental chromosomes is normal. Thus the number of healthy transmitters can be much higher than the above figures. In the case of mucoviscidosis, for example, where the condition is handed down recessively from one generation to the next, the morbid gene is carried by about one person in 20.

Alongside the equation "1 defective gene = 1 disease," there are also relatively frequent situations in which the changes affect not one gene, but several simultaneously. Most often, these alterations determine predispositions to the disease, rather than causing it directly. The action of environmental factors, for example, a viral infection, must also be taken into account. The number of people suffering from polygenic conditions is relatively high. These diseases include insulin-dependent diabetes, arteriosclerosis, hypertension and some forms of depression. There is often a statistical correlation between the onset of a polygenic condition and the alteration of a histocompatibility locus.

Where does genetic diagnosis stand at present? Efforts have long been under way to foresee the precursors of disease, either in the fetus or in the newborn child, or even in some cases in adults when they come from a high-risk family. In the past, such diagnosis was based on measurements of biochemical or immunochemical parameters, or on examinations of the karyotype (the individual's chromosome formula). For example, people affected with Duchenne's disease have an abnormally high rate of creatine kinase, which can thus serve as an indicator. Another well-known example is that of Down syndrome, which involves an additional chromosome 21.

The progress in molecular genetics means that copies or significant fragments of genes, known as genetic probes, are now available. Sometimes these are fragments obtained through chemical synthesis processes involving automated operations performed by machines "programmed" accordingly (oligonucleotide synthesizers). Most often, however, they are true copies of messenger RNA (cDNA) produced by converting them using reverse transcriptase (transcriptase is the enzyme which catalyzes the transcription of genes into RNA; reverse transcriptase enables the transition to be made from RNA to DNA). Most frequently, these probes are labeled by means of a radioactive tracer, or combined with amplifying reagents. They can thus be detected after they have been made to interact with the gene being analyzed for possible changes. This is done by hybridizing the probe with a preparation of DNA extracted from cell samples taken from the patient.

When there is no alteration in the gene under study, the hybrid form is "perfect," thus making it completely resistant to nucleases, the enzymes capable of splitting nucleic acids. If the contrary is the case, and the hybrid is imperfect, the anomalies can be detected by electrophoresis or digestion tests. For example, hybridization with a "hemoglobin probe" will make it possible to determine whether or not the patient shows deletions in the globin genes, alterations in the regulatory mechanisms (thalassemias) or localized mutations (sickle cell anemia).

These techniques do not require any complicated apparatus. However, obtaining the probes is sometimes a delicate operation. It is nevertheless foreseeable that, in the near future, it will be possible to use these processes in a doctor's office or laboratory. In 1974 two American doctors, Kan and Dozy, first used genetic probe hybridization techniques in human pathology to look precisely for alterations capable of affecting the hemoglobin genes. Since then, fifteen or so of the most common hereditary diseases have come to benefit from this methodology.

As already indicated, these genetic tests can be carried out either on the fetus (prenatal diagnosis) or after birth (preclinical diagnosis), or again in adults in order to identify carriers. As compared with the other testing methods, prenatal diagnosis based on the use of genetic probes offers the advantage of greater reliability, generally over 90%. In addition, it can be conducted at a relatively early stage of fetal development, from the eleventh week of pregnancy onwards. At this time, samples are taken of the DNA present in the cells of the trophoblast, that is, the outer epithelium, outside the embryo, from which the egg derives its nutrition. The number of prenatal diagnoses has increased dramatically since the beginning of the 1970s.

The trend towards this systematic screening seems to be gaining momentum. It is motivated not only by the concern to counsel couples as to the desirability or not of having children when one of the spouses belongs to a high-risk family, but still more by the aim of developing a genuine predictive medicine. The aim of this medicine may be simply to draw attention to the consequences of a disease for the child that is to be born, when no therapy is available and the disease may have a fatal outcome. When even a partial therapy is possible, however, it may involve instituting certain forms of treatment, or at least more regular medical monitoring, if necessary immediately following birth.

However, the difficulties should not be underestimated: most often, these practices amount to no more than a kind of therapeutic uncoupling, to the degree that no remedy has as yet been found for the majority of the diseases in question. The problem here, which presents a serious challenge for our society, is thus one of "knowledge without power."

The possible legal and moral consequences of an expansion in genetic diagnosis are readily apparent. For example, would not the parents of a child born abnormal be justified in suing the physician who failed to have recourse to genetic diagnosis? The problem is all the more serious in that American and French surveys clearly show that, in these matters, information is not equally available at all levels of society: genetic screening is resorted to much more frequently among people belonging to the liberal professions and the well-off classes than among the rest of society. Is not the concept of loss of opportunity, which nowadays is so often invoked against doctors on the basis of the principle of the right to health, likely to be extended to failures in genetic information as well? I for my part am convinced that it is.

Conversely, it must be acknowledged that, a certain portion of the population will prefer to remain in ignorance rather than to bear the psychological weight of the genetic burden. The case of Huntington's chorea, an extremely rare disease but a terrible one, in that it appears after the age of forty and exposes the victim to an anguishing progressive neurological breakdown resulting in death, has often been raised by ethics committees; should not the right to ignorance be respected in this case? The same comment applies, moreover, to Alzheimer's disease, the locus of which has not yet been identified. It may be carried by chromosome 21, and the changes may be close to those that underlie Down syndrome. As we know, this disease is manifested in impairment of memory from the age of forty or fifty onwards, followed by a loss of behavioral autonomy, and by the main syndromes of senile dementia.

There can be all the less doubt that the focus will increasingly be on genetic diagnosis of precursors in that, as we are about to see, progress is also being made towards full knowledge of the human genome. Apart from these extremely extensive operations, aimed at comprehensively sequencing all the chromosomal DNA, mention must also be made of another approach: this, too, involves an ambitious program, already in large part implemented, the aim of which is to study genetic polymorphism in humans.

This program is being conducted at the international level in the Center for the Study of Human Polymorphism (CEPH) in Paris, under the leadership of Professor Dausset. Some forty families have been selected, and the aim is to come up with a sufficient number of probes—several hundred of them—corresponding to restriction markers, so as to facilitate the matching up with chromosomes of genes that have not yet been mapped. This program has, for the most part, been completed. A considerable number of "tools" are now available, thanks to the joint efforts of European and American researchers. It will shortly be possible to distribute restriction probes, as they are known, free of charge to laboratories that request them.

In the light of these various efforts, two related problems will arise. The first concerns the confidentiality of these operations, and above all of the resulting data; by analyzing people's genes at the molecular level, we are reaching deeper and deeper into their innermost beings. Who will have access to the genetic registers, apart from researchers and physicians? The second problem relates to genetic ostracism and its systematized form, eugenics. The term ostracism is used here to mean the labeling of an entire category of individuals with a view to segregating them. This may take various forms. Apart from exhaustive genetic analysis of chromosomes, a technique may be used which has recently become famous as a result of its successful application in police work in Great Britain. Here, genes are of interest not for the purpose of seeking out diseases, seen as possible taints, but in order to establish beyond doubt, through overall genetic analysis, that a given cell does indeed come from a given individual by detecting what has been called "genetic footprints." Identification of paternity is one aspect of this approach, but it is gradually being turned to more general use in assisting immigration authorities, for example in the United States.

The main risk seems to lie in the extension of this record-keeping procedure to all citizens. Administrations would then have, with the genetic map, an infallible means of identification which would be much more revealing than the use of fingerprints, for the latter afford no possibility of extrapolating data regarding the physiological individuality and potentialities of the person concerned, or of the risk factors that can be attributed to him or her.

THE TEMPTATION OF EUGENICS

It would be too easy here to fall into Manicheism. After all, are not everyone's private affairs already being more or less analyzed in detail by computers? But although laws exist guaranteeing the confidentiality of information, the problem here is of a different dimension. Eugenics is not far off. The temptation will be strong, once accurate registers of individuals' genetic formulas are available and lists can be drawn up of their histocompatibility genes, their risk factors and mutations bearing the stamp of hereditary diseases, to use them for purposes which go beyond preventive medicine (insurance companies, employers, etc.). Again, it would not be difficult to legislate on the subject, and various ethics committees have already pronounced themselves on this aspect. However, a kind of imperceptible slide towards eugenic systematization is starting to develop, and this it seems to me is a matter for serious concern.*

When there is a grave threat to the life of an as yet unborn child, recourse to genetic counseling, most often followed by abortion, may appear to some people as a clear act of eugenics, whereas others will see it as an essentially humanitarian act. Our aim is not so much to

---

* The Office of Technology Assessment (OTA) is launching a major survey of the abusive utilization, which is already taking place with disturbing frequency, of data deriving from the genetic analyses of job applicants and employees now being conducted by various industrial firms.

take sides as to emphasize that, if this is eugenics, each partner can at least react to it in the light of his or her own sensitivity, personal ethics and beliefs. But if the study of human genes, that is, their comparative analysis at the individual level, becomes widespread, care must be taken to avoid a kind of psychosocial change of course involving a more or less conscious transition from the concept of "hereditary trait with morbid or lethal effects" to that of "implied deviant hereditary trait"—deviant, that is to say, from the "norm."

Whether or not there is a scientific and medical basis for this alleged "norm" hardly changes the problem, of course, the moment preventive selection becomes *compulsory*. In the face of possible attempts to match individuals' genomes to certain standardized reference points, it should be recalled that in humans, as in any species, it is polymorphism and not genetic uniformity which is the rule.

Dr. Kourilsky has provided some extremely interesting quantifications in this respect: "Let us assume an average of one difference per 1,000 bases between two individuals, which is probably an underestimate. This would give three million differences, at least, between two human genomes containing some 3 billion base pairs." (*Les artisans de l'hérédité* [The artisans of heredity], Odile Jacob, 1987). To speak of a standard gene, or a reference gene, thus makes no sense.

It must also be recognized that the dividing line between a mutated gene whose product is defective and a mutated gene whose product retains normal or quasi-normal function is sometimes extremely slender. There is a serious risk of an ill-considered extension of the practice of prenatal diagnosis occurring, an extension aimed at a measure of standardization of the genetic heritage within a given society.

Caution, indeed uncompromising resistance, is therefore essential if we wish to prevent systematic molecular eugenics from becoming established. The world will require a great deal of wisdom in order to be able on the one hand to stay on course towards scientific progress and on the other hand to prevent its appropriation for sociopolitical or ideological ends. I already mentioned Dr. Muller-Hill's indictment, but reference may also be made to a magazine article (Marc Rocheman, *Biofutur*, February 1988) where the author recalls a number of events which, in the light of history, afford a sad illustration of what could be termed positive eugenics.

These included the mass sterilization campaigns conducted between 1920 and 1930 in some thirty American states, directed against alcoholics, the abnormal and the feeble-minded, as well as the Nazi experiments and the search for a final solution. We may also note—an indication that eugenics, alas, knows no bounds—the peremptory statements by Charles Richet and Alexis Carell, two Frenchmen, both Nobel Prize winners, passionately preaching the elimination of the abnormal. And the author concludes that "these trends still exist in society, and have permeated the minds of part of the public as well as of lawyers and scientists, who are 'people like everyone else.' AIDS is currently providing fairly crude revelations of these trends: tattooing of seropositives on the genitals (proposed by a Swiss doctor); isolation of patients in 'sidatoriums' [from SIDA, the French acronym for AIDS] (proposed by a French politician); bars to the exercise of professions, etc. The new power of biotechnologies over life, coupled with these old but still current trends, lends a special dimension to the current ethical debate."

But I should like for a moment to leave the field of the history of public health, and even of genetics.

The fact of being a biologist and having a passion for one's work, as I have, does not prevent a certain sense of oppression, at times, in the face of extrapolations from this science which may appear to be normative, classificatory, or simply capable of serving as points of reference. The fact that we now have a better understanding of human biology does not mean that there is any need for us to increasingly reduce people to their strictly biological dimension. Thus, on rereading *La tentation eugénique* [The eugenic temptation], an article by Albert Jacquard, I am inclined to accept his reasoning: "What would eugenics be used for?" he writes, "To develop better genes, a better environment? What we need to develop is our own ability to develop ourselves. Would it not be a fine objective for a society, rather than improving genes in unknown ways and directions, to improve the capacity to make use of this magnificent gift nature has given us and which we are currently preventing from manifesting itself?" (*Les Cahiers du MURS*, October 1984, No. 1).

In the same frame of mind, we might quote the philosopher Georges Canguilhem, for whom any talk of norms is in the first place a discussion of departure from those norms. Are we faced, once again, with an "irreversible" advance of science which, because it will sooner or later become normative, will by the same token claim to be "moral" in Durkheim's sense of the term?

Another example, and a highly topical one, of the intrusion of biology into culture relates to the control of reproduction. Professor Jean Bernard reminds us that this is something which goes back to the earliest times. In ancient Rome, for example, a doctor advised women to hold their breath and sneeze at the decisive moment in order to avoid becoming pregnant. So much has been said and written on the problem of human reproduction that I may be forgiven, here even more than elsewhere, for not attempting to be exhaustive.

The modification of human embryos by the use of transgenosis techniques, though still very much in the realm of hypothesis, brings us to the verge of what Henri Atlan, in *Le jaillissement des biotechnologies* [The flood of biotechnologies] calls "a kind of genuine positive eugenics." By this he means an attempt to "manufacture" a being endowed with the qualities looked up to by the majority, as opposed to negative eugenics, which involves "eliminating" the part of our natural endowment that is deemed not to come up to the standards we impose. Viewed in this light, the technologies of reproduction which, thanks to the inexhaustible inventiveness of semantics, are sometimes dubbed "medically assisted reproduction" would form a fairly logical extension of genetic engineering techniques. But I shall touch on this subject only briefly, primarily to stress its cultural resonances.

Medically assisted reproduction techniques began with artificial insemination. This practice, which is extremely widespread in agronomy, animal husbandry and biotechnology, has played a major role in the reproduction of various animal species. Its application to the human species began in 1958, when the idea was put forward of advising young men threatened with sterility as a result of chemotherapy to resort to sperm conservation. The matter became somewhat more complex from the time when, in response to accidental sterility in the male partner of a couple, the same type of procedure was employed using the sperm of anonymous donors. The problem that then arose was, essentially, one of dissociation between the act of love and the act of reproduction, as Jean Bernard so aptly expresses it. Anonymity of the genetic father can induce in the couple, even more than in the child, psychic disturbances of which the specialists are well aware.

The most familiar fact of the new practices, and one which has given rise to the most impassioned debate and severely tested the majority of ethics committees, in particular, relates to *in vitro* fertilization, that is, fertilization of the ovum by the spermatozoon outside the female organism. The first "test tube baby" was Louise Brown, the little girl born in July 1978 in a clinic near Cambridge established by Edwards and Steptoe. Other "test tube babies" were subsequently born in France, Australia and Germany. Medicine does indeed render a great service here by reducing the misfortune of sterility. Most often, the aim is to make up for an obstruction of the tubes. But one of the arguable consequences of this practice cannot fail to arouse concern. Given the relatively large number of failures, the doctor fertilizes a number of ova in the laboratory, and of course only reimplants one of them. Biologists and obstetricians thus have on their hands a relatively large number of frozen human embryos that they do not know what to do with, whether to destroy them or keep them.

And of course, the problem becomes more complicated when a surrogate mother is used. A child conceived in this way literally has two mothers, and the lengthy debates to which this unusual, but not unique, situation has given rise are well known. Again, we are dealing here with scenarios which, in part at least, involve a physiologically natural development process. But two more fantastic situations can already be glimpsed on the horizon. There is every possibility that they could become a reality in the course of the 21st century.

The first of these would be the development of a child *entirely* outside the female body. As of now, this operation is impossible, since beyond a few divisions (the 16-cell stage), the fertilized oocyte can no longer be reimplanted, and complete embryogenesis is impossible *in vitro*.

The second relates to the possibility of transgenosis in one of the surplus oocytes mentioned above. It is quite conceivable that a genetic diagnosis could be established on one of the ova released by artificial stimulation. This ovum would subsequently be disposed of, but it would serve to identify the extent of any possible genetic lesion. The genetic correction would then be performed on another ovum, before reimplanting it in the mother's uterus.

At present, of course, any genetic transfer, whether it involves somatic or gametic cells, to an individual or fetus is prohibited. Likewise, it is illegal to carry out experiments of this kind on ova, whether fertilized or not. On the other hand, they are authorized on human somatic cells kept in culture, on condition that these cells are not reimplanted. Here, as in the beginnings of genetic engineering, when biologists accepted the idea of a moratorium, international ethics has expressed the will not to go further, or rather to set precise limits on how much further to go.

The problem is in fact the same as that faced by genetic therapy: in seeking to become angels, we do not wish to run the risk of turning into demons. And yet, the practice of somatic transgenosis, using for example bone-marrow transplants on which genetic transformations had been performed *in vitro*, should be the logical next step if we want to be able to control severe hereditary diseases effectively some day. Unless we resort to abortion, it is indeed difficult to see how this step can be avoided, if the disease cannot be circumvented by the usual treatments, particularly chemotherapy. Unless society decides otherwise—either because it deems the cost of the activity too high or prefers to let things take their course.

If, as many biologists think, genetic therapy at the level of the soma will mark a decisive turning point in tomorrow's medicine, on a par with the use of organ grafts, this will call for perfect mastery of gene transfer techniques*; as yet, however, we do not know with certainty how to direct the foreign, corrector, gene to a precise location on the chromosome. There is thus the risk of damaging a normal gene in an attempt to repair the defective one.

As Henri Atlan very wisely emphasizes, the introduction of biotechnological practices, primarily those of transgenosis to fertilized human ova, does considerably strengthen our ability to see ourselves as a biochemical system reduced to molecular interactions, even though at the outset this was done in response to therapeutic demand.

---

* At the time these lines were written, it was still not known whether the commission responsible for examining all experimental protocols involving the use of genetic engineering techniques in the United States had authorized the use of genetically manipulated human cells in the treatment of certain cancers. It appears that the experiment is to be conducted in two phases: initially, an effort would be made to label lymphocyte cells (genetically) before reinjecting them in order to monitor them in the organism. In the second phase, the therapeutic effort proper would be made by recombining the interleukin-2 gene with the lymphocyte cells to be reimplanted in patients. This information thus strengthens and updates our argument. There is as yet nothing to indicate, however, that we are now entering the age of genetic therapy. This will naturally depend in large measure on the outcome of these experiments.

IV

TOWARDS A
PLANETARY
BIOLOGY?

Since the end of the Second World War, the public has become familiar with the concept of "large-scale" projects, such as the development of high-speed trains, the space shuttle, new satellites, fast breeder reactors (Superphoenix), supercomputers and vast particle accelerators.

Physics is the predominant science here, and the public expenditure is considerable. Given the enormous size of the budgets required to carry out these immense programs successfully, over time spans of between five and ten years, the major powers have established cooperative links at the continental or intercontinental level. The exploration of space is certainly the best-known example.

One may well consider that this state of affairs has helped rank physics in a position of cultural predominance, so that it has long exemplified modern science. The life sciences, although they too are sciences of nature, have long occupied only a secondary, if not subordinate, place in society's concept of science.

And yet, one of the newest features of the new technological civilization coming into being before our very eyes comprises, as this book shows, the increasing role played by the life sciences in this concept. Biotechnologies, as we have seen, now rank high on the honor roll of advanced technologies, and the tragic advent of AIDS, in a society which believed itself free from major infectious diseases, has lent biological research new topicality and urgency.

But there is also a new element contributing to this rise, or if you will, this promotion: biology is beginning (whether rightly or wrongly matters little) to provide the inspiration, like physics, for large-scale and expensive projects.

To conclude, I should like to refer to two weighty examples which, other things being equal, exhibit two common characteristics: on the one hand, both projects are based on a universalist conception of biology—a conception which holds that this science can, given its current dynamic state, overthrow some of the foundations of society, for the benefit of all; and, on the other hand, implementation of these projects is apt to confer solid political and economic advantages on the powers that launch them.

KNOWLEDGE OF THE HUMAN GENOME

The first of these projects is known as the sequencing of the human genome; it was initially conceived and has to date been fairly broadly developed in the United States. The second has just recently become the focus of a vast international mobilization effort; it was proposed and, as we shall see, strongly promoted by the Japanese, who are coordinating its implementation, and it is inspired by the concept of a renewed bionics. This is the "Human Frontiers" program.

For a number of years now, many biologists have been taking an interest in the human genome. That this interest has come so late may be something of a surprise. But it should not be forgotten that the hereditary material imbedded in the nuclei of our cells is one of the most complex, in terms of its DNA content and of the large number of chromosomes—twenty-three pairs—through which Watson and Crick's immense double helix is coiled. For a long time, biologists had enough on their plate with the chromosome map of *Escherichia coli*: the bacterium has only a single circular chromosome, but this chromosome, remember, contains no fewer than two thousand genes. Small wonder, then, that they had their work cut out for them just with analyzing *Escherichia coli*.

It was thus not until the beginning of the 1960s that the assault on the problem of human chromosomes began. At that time, no physicochemical method was available for fractionating them, and a stratagem was thus resorted to which stemmed from the data of experiments on hybridization of somatic cells. Since Barski's work, it had been known that two cells from different species can artificially be compelled to fuse together: for example, a somatic hybrid between a human fibroblast and a mouse hepatic cell (hepatocyte) can be produced, with the nuclei from the two cells also coming together to form a single entity, and the hybrid cell is capable of producing progeny when cultivated *in vitro*.

A remarkable, and still poorly understood, fact is that the human chromosomes are gradually eliminated from this hybrid nucleus, whereas the mouse chromosomes persist.

On examining the daughter cells, nuclei can therefore be found whose chromosomes are entirely of the mouse type, and which also enclose a varying number of human chromosomes, sometimes even just a single one. It can thus be deduced, by a kind of subtraction, that a given property expressed by the hybrid cell must be attributed to a given specific chromosome of human origin, and this makes it possible to establish step by step the location of the determinant for a specific human phenotype.

The first human chromosome maps were plotted using this technique. Then, gradually, the arsenal of physical and physicochemical methods became more extensive. It is now possible to sort the chromosomes, using special machines capable of assessing the fluorescent state of given chromosomes after interaction with an appropriate reagent, etc. Above all, however, since genetic probes and DNA splitting techniques using restriction enzymes have become available, it is possible to subject the hereditary material of a given chromosome, or of an entire nucleus, to treatments which transform it into a multitude of fragments. These are then separated by gel electrophoresis, that is, under the impact of powerful electric fields, and what is known as the physical map of the genome is drawn up.

Study of complex genomes is all the rage, and has proven remarkably effective. It is based on the detection within chromosomes of the minor chemical changes that result from mutations—mutations which mean that, genetically speaking, no individual is like any other, even if they have the same parents. Scientists have been able to calculate that this genetic polymorphism, the exact extent of which was until very recently not capable of being fully grasped for lack of sufficiently precise techniques, is reflected by a change in a pair of DNA bases with a statistical frequency of 1 per 1,000, with the result that between two brothers, for example, these distinctive minichanges, which are neutral in terms of their physiological consequences, affect a few million chemical sites on the DNA. These discrete changes are experimentally detectable only as a result of the modifications they bring about in the splitting patterns due to restriction enzymes: the fragments produced differ very substantially in length. This is what is known as "restriction-fragment-length polymorphism" (or RFLP).

Since genetic probes exist for these RFLP sites, an excellent detection method is available which in principle facilitates the location of any hereditary trait on the chromosomes. This is thus an incredibly fertile form of indirect molecular genetics. Nor should we forget that the old and trusty technique of analyzing what are known as "pedigrees," or the study of the distribution of a hereditary trait in a family's offspring, also yields a great deal of information. For example, it can be used to determine whether an altered gene is carried by a sex chromosome or not, whether the mutation is dominant or recessive, etc. When the mutation in question is accompanied by cytological changes that are easily observable under a microscope (breaks, translocations), the target gene of the mutation can also be assigned to the specific chromosomes containing these alterations.

Through a combination of these various approaches, it has thus been possible to locate a considerable number of genes on the human chromosome—about 1,500, judging by the latest estimates of the International Congress on Human Genetics held in Paris in 1987. But there is still a long way to go! This can be seen from a few truly impressive figures. Remember that the total complement of human chromosomes housed within a cell nucleus contains the equivalent of three billion pairs of nucleic bases. Assuming, roughly speaking, that each gene codes on average for one protein with a mass of 40,000, one gene can be assigned a length corresponding to 1,200 pairs of bases.

The nucleus of the human cell should then contain 2.5 million genes. In fact, however, given that the greater part of the DNA has no coding properties (consisting of repeat sequences, etc.), and in the light of a whole series of independent measurements, biologists are in agreement that there are barely more than 50,000 to 100,000 distinct functional genes. Even so, with only 1,500 genes located, the task remains an immense one if we want to know the locus, and still more the nature, of *all* other human genes.

It should also be added that the chemical sequence has been precisely established, after cloning, for only two to three hundred of the genes thus far located. And knowledge of this sequence is extremely important, because on the basis of the combinations of the genetic code it is easy to deduce from the chemical sequence of a gene the sequence of the corresponding protein. In turn, knowledge of the protein most often provides extremely precise information regarding the mode of action of the gene in question; the protein can be, for example, a hormone, an enzyme, an element in the cell architecture, or a regulatory factor.

Against this background, American researchers a few years ago launched the idea of an international program that would mobilize a very large number of researchers and considerable financial resources and would involve the design of new automatic measuring instruments in order to draw up the complete sequence of the human genome.

It may be estimated that, using present-day techniques, a project of this kind would keep 150 to 300 research workers busy for some twenty years! Obviously, this estimate will have to be revised downwards if, as is hoped, new automatic sequencers can be built. American, Japanese and European designers are at work. It is hoped that some of these machines will have the capacity to read the sequence of several dozen or even several hundred thousand pairs of bases a day. But that point has not yet been reached. And the cost of the operation will be between one and three billion dollars.

An undertaking of such magnitude, in terms of both its financial aspects and the scientific and logistical skills involved, has not failed to arouse a great deal of debate within the scientific communities in the countries concerned, primarily, of course, in the United States where the project originated. A bill was introduced to provide Federal funding for the program. Four chromosomes (8, 16, 19 and 21) have been selected for priority sequencing. The Department of Energy has been brought in and has designated the Lawrence Berkeley Laboratory and the Los Alamos Laboratories as supporting structures for this vast undertaking. The financing is drawn from both the Federal authorities and various scientific foundations such as the Howard Hughes Foundation. The program being conducted under the responsibility of the National Institutes of Health is also an impressive one.

Neither the Japanese nor the Europeans have remained indifferent to this challenge. The former are staking a great deal, as we have seen, on the construction of giant sequencers, drawing on their highly developed electronics industry. The latter, although it appears that they have not as yet arrived at any multilateral decision at the Community level, are holding frequent consultations and meetings and, with the help of the powerful DG XII,* beginning to marshal their forces.

In Europe, as elsewhere, opinions remain divided. Would it not be better, if the aim is to understand a complete genetic algorithm, to attack genomes that are smaller in size? If so, would not yeast, *Drosophila*, the nematode worm or the plant *Arabidopsis* seem more appropriate? A commitment to sequencing the human genome will not only cost a great deal, but risks diverting biologists, for a time, from projects which would perhaps be more useful, in the neurosciences for example.

Of course, the project's supporters reply, but nothing will really replace knowledge of the human genome in diagnosing and combating disease, and of the 3,000 hereditarily transmissible diseases that have been listed, a good

---

* The Directorate-General for Science, Research and Development, which is responsible for helping support or launch technological research and development programs within the European Community.

number affect the nervous system. And only a few dozen or so of these have been precisely located in terms of the corresponding genetic alterations. Other researchers place more emphasis on the considerable benefits that would accrue from being able to compare the complete genetic map of humans with that of other animals, particularly the higher primates, in order to improve our knowledge of the branchings in the course of evolution. Lastly, some biologists, such as Professor Walter Gilbert, one of the major forces behind the American project, go even further. At a recent international symposium in Rome on the ethical issues relating to the study of the human genome, he did not hesitate to tell his audience that the complete sequencing project marks a decisive turning point not only in predictive medicine, but also in molecular biology.

To know all human coded genetic information is to give us access, at least in a first approximation, to an array of data on the mechanisms involved in human development, physiology, even behavior. If the industrialized world undertakes such a project, data processors, among others, will become heavily involved with biologists in storing, sorting and processing the data. Converting the linear information of the sequences into three-dimensional protein structures, manufacturing a host of antibodies against the thousands of unknown proteins, trying to localize these proteins in the cell and compare their level of expression in the course of normal and pathological development—all this will constitute an enormous amount of work. Yes, Gilbert is no doubt right: this would indeed mark the advent of a new era in molecular biology, the era of *Homo sapiens*. Numerous teams would be mobilized, and that alone could make the situation one of unprecedented sociological impact.

But the ethical implications and consequences of this need to be seriously examined. For example, will the data on humans be directly accessible to all, or exploited commercially? Is this truly the most interesting and most useful project in the life sciences? These are two questions which are most likely to give rise to extensive international discussion.

HUMAN FRONTIERS

In 1984, I was at the charming site of Hakone, on the shores of a lake close to Mount Fuji. But I was not there to admire the scenery. I was attending an international gathering, together with my colleagues Jacques Ruffié and Michel Serres. The Prime Minister of Japan, Mr. Nakasone, had convened a highly formal meeting of a panel of scientists, jurists, philosophers and theologians from the seven most highly industrialized countries. Their agenda was to debate, in very broad terms, the ethical problems raised by the recent developments in biology and its applications.

A year earlier, François Mitterrand, returning from a trip to Williamsburg to attend one of the famous big-power summits, had invited me to lunch along with a few other people. Commenting on his impressions, he said he had been surprised at the attitude of the Japanese Prime Minister: far from confining himself to the customary dialectics on trade or the transistor and electronics markets,

Mr. Nakasone had expressed an interest in questions relating to biology and medicine. He was thinking of a major meeting on cancer. François Mitterrand was astonished at this statement of interest in a field which had not until then seemed to attract Japan's attention. It appeared that Mr. Nakasone, who had for a time been in charge of public health, now wanted to take a line which was more of a departure from his country's technical and consumerist image.

The envisaged big meeting on cancer never took place. Instead, Japan decided to favor a much broader approach to biological problems. Hence the first symposium on bioethics attended by the wealthiest countries, which brought together Americans, Canadians, Japanese and Europeans, and set a pattern since then never broken of an annual meeting, each time in a different country. This was for Japan a significant move on the chessboard of politics. That is why those who have had the privilege of visiting Japan frequently and maintaining regular contacts with its government were only moderately surprised to learn, at the end of 1985, that Japan was proposing to launch a vast project inspired by biomedicine, a project which from the outset was christened "Human Frontiers."

What exactly did this involve? It called for participation in an international undertaking, centered in Japan, whose aim was nothing less than to rebuild a society made up of technical components, but inspired this time by biology. It expressed a new philosophy of the modern age, based no longer on rather vague reference to nature, but on the knowledge of cellular and molecular biology acquired in recent decades.

The underlying concept was stated in the preface to the Japanese document: our technological society has entered a phase of irreversibility, the irreversibility of modern-day logic—yet a menacing irreversibility, at a cost which already includes many traumatic events repeatedly conjured up since the end of the last war: pollution, depletion of natural resources, energy overconsumption, technological risks, etc. The solution would involve a renewal not merely of our technologies but of our entire methodology, drawing its inspiration from the performance (and the "refinements") of the biological world, which over the course of its evolution had arrived at a peak of efficiency, integration and harmony. This project, clearly, derives from a kind of updated Rousseauism. The approach it advocates is inspired by a renewed form of bionics, a science about which we now need to say a few words.

The term bionics was introduced, some thirty years ago, by an American researcher, J.E. Steele, to describe research into cybernetics, that is, research aimed at analyzing and above all simulating, through machines, the functions characteristic of living beings. In fact, the term "bionics" is a contraction of the words biology and electronics.

One of the characteristics of living organisms which from the very outset presents a challenge for cyberneticians is its complexity: the human cortex contains no fewer than ten billion neurons. No less remarkable are its faculties of adaptation and the economy of means it employs. For a long time, with the help of mathematics and the use of electronic circuits, researchers sought to simulate the nervous system as well as the sensory and effector organs; they also embarked on a study of the analysis of forms (pattern recognition). As early as 1933, N. Rashevsky had imagined the possibility of depicting the functioning of a neuron schematically by means of a decision element capable of existing in a binary state, able to pass from a state of rest (no: 0) to a state of excitation (yes: 1) as soon as the sum of afferent stimuli exceeded a certain threshold.

This led to the concept of the neuron network, an assembly of formal elements distinct from the cell neurons. These elements are linked by connections, and interneuronal transmission can take the form of facilitation or inhibition. This formal structure lends itself to interesting mathematical developments, showing that an assembly so conceived, although composed solely of elements with a binary potential, can give rise to the possibility of performing complex operations; in particular, it can simulate memory, and adapt to imposed situations. Among the many applications of bionics, or at least those which initially became apparent, we might mention medical diagnosis, meteorology and any situation in which groups of morphological facts need to be analyzed.

Other research (Haldan Keffer Hartline and William Reichardt) related to vision in invertebrates (the photoreceptors of *Limulus*, an arthropod) or vertebrates (the work of Lettwin, Matturana, McCulloch, W. Pitts, *et al.*, on the pigeon retina), to the mechanisms of motor control and the execution of movement by muscle fibers and, as already mentioned, the analysis of visual and sound forms.

The Human Frontiers project entails too many subdivisions for us to analyze exhaustively here. Such an analysis would moreover involve several "grids": thematic, disciplinary, technological, etc. The general objective is to obtain an understanding of the whole range of biological and psychophysiological mechanisms, starting with the more "molecular" and moving towards the more "cortical," so to speak.

The intention is to cover both molecular interactions, the mechanisms of genetic expression and cell transduction, the major bioenergetic processes (with interest focusing to a much greater extent on animal models and humans than on the plant kingdom), and the higher cognitive functions, etc. Consideration was given to the practical relevance of each of the relating lines of research—applications of interest in pharmacology and instrumentation; development of new tracers, new sequencers, non-invasive instruments for use in clinical anatomical investigations, etc.

One can see why, given the breadth of this program and its ambitious nature, the first reactions by the major powers were reserved, to say the least. It was pointed out to the Japanese authorities (MITI, AST, embassies) that Human Frontiers was likely to find itself competing with a whole series of international programs already under way whose implementation might be slowed by such a major effort in the directions proposed. The Americans consulted (particularly Professor J. Lederberg) recommended a complete change of content, advocating themes more closely related to the specific problems of the Third World. As for the various diplomatic authorities, they exhibited a certain feeling of panic, to say the least, not only at the program itself, but at the avalanche of Japanese proposals that ensued.

If we add to this the fact that initially the Japanese scientific community and still more the Ministry of Finance did not seem particularly enthusiastic about MITI's proposals, there were grounds for fearing that the project might simply vanish without a trace, particularly since both Americans and Europeans were barely concealing their fears that Japan might take advantage of this large-scale international scientific cooperation to attract the best minds so as to make up for its lag in certain areas of basic research, particularly biology. Aware of these reservations, the Japanese authorities decided, towards the end of Mr. Nakasone's mandate and then with the entry into office of the new Prime Minister, Mr. Takeshita, to change their strategy.

A "feasibility" committee chaired by Professor Okamoto, President of Kyoto University, was established. Its Japanese members embarked on a series of visits to various world capitals, emphasizing that the actual agenda of the project would be decided later by an assembly of international scientific experts, but would draw on two main themes: study of molecular interactions and the higher cognitive functions. They also specified that Japan would provide financing for the principal operations (exchanges of researchers, missions, joint programs, symposiums, etc.) for the first two years of the program's implementation. In this way, they were able to restore a climate of confidence.

The project was approved in principle at the Toronto summit of the major economic powers. The machine has thus been set in motion. If it succeeds in achieving its desired end, it will mean that the Empire of the Rising Sun now means to become, scientifically but culturally and economically as well, one of the major world centers for biology, biotechnology, and even bionics.

Whatever interpretation may be given to these "great schemes," and whatever the fate of these major projects may be, it seems to me obvious that biology and biologists are now at the core of the most serious political stakes. Now that politicians are seeking to draw all practical consequences—economic, social and human—from the accelerated production of new knowledge, it is indeed the fate of our civilization that is at stake.

Such, then, is today's biology: after a long struggle to free itself from the tutelage of physics and chemistry, as well as of philosophy and religion, it has had to become more precise in terms of its objectives—to understand how life is structured and functions—and in terms of its methods. It has had to cast off at long last the yoke of an all-powerful medicine, at the price of a covenant with its former overlords, physics and chemistry. But now it in turn is on its way to assuming tutelage over others. Strengthened by its successes and driven by its ambition, it seems indeed to be becoming the prime point of convergence of the "hard" sciences: mathematics, physics and chemistry. At the very least, no one will dispute that it has now taken shape as an integrated science, borrowing widely from computer science and thermodynamics, as it once did from mechanics and cybernetics, to elaborate its models.

Such is the situation of biologists today: Our society, whether consciously or not, is turning towards them as if its fate depended in large measure on the progress to be made.

Whether in combating hunger or disease, in reducing pollution, in finding ways of warding off the traumas of life in the megalopolis, or in regulating world growth, a great deal is expected of biologists, who have been elevated to the rank of the great sages of modern times.

Aristotle reserved pride of place, among all the functions of living beings, to generation because of the extraordinary metaphysical importance which attaches to the permanence of form in species. So firmly rooted in us is this idea that genetics, the science of heredity, has always aroused a measure of anxiety and the public has in these areas often and for a long time confused the cognitive approach and the hands-on approach.

Agriculture and animal husbandry have indeed made us familiar with the idea of deliberately changing *natura naturans*, whether by craft or by science, but this has been more a matter of utilitarian manipulation than of a deliberate attempt to change living matter and its characteristics. As we have seen, recombinant DNA techniques have expanded the potential for acting on species to the point where biology goes to the very heart of the great questions facing our societies.

Thus, through its technological and scientific, as well as its moral and philosophical dimensions, "biology," as Bruno Ribes writes, "is faced with increasingly urgent pressures and demands from outside, which remove it further and further from its natural bent."

Being a biologist today is an exciting and difficult task. Exciting because the social and cultural challenge is on no less a scale than the striking advances the discipline has made. Difficult because, between the siren song, the "Durkheimian" fear of the moralists, and the demands of scientific rigor, biologists do not always know what attitude to adopt: one of retreat into their ivory towers, or one of openness? How are they to keep cool heads and maintain scrupulous respect for the rules and criteria of the scientific approach while fulfilling as responsible citizens, indeed as fellow human beings, the expectations of a world which wants to see things move ahead at full speed? More important still, how are they to avoid biology serving—as history, just a few decades ago, taught us that it can—as a pretext for the most deadly ideologies?

I hope these few pages will help everyone, biologists and nonbiologists alike, to approach these awesome questions in the right way.

BIBLIOGRAPHY

ATLAN, H., *A tort et à raison, incertitude de la science et du mythe* [Rightly and wrongly—the uncertainty of science and myth*], Le Seuil, 1986.

BERNARD, J. and A. JACQUARD, *Cahier du MURS*, No. 1, October 1984.

BUICAN, D., *La révolution de l'évolution* [The revolution in evolution*], PUF, 1989.

DAGOGNET, F., *La maîtrise du vivant* [The mastery of life*], coll. "Histoire et philosophie des sciences," Hachette, 1988.

DARBON, P. and J. ROBIN (coordinated by), *Le jaillissement des biotechnologies* [The flood of biotechnologies*], Fayard-Fondation Diderot, 1987.

GROS, F., *Les secrets du gène* [The secrets of the gene*], Odile Jacob, 1986.

KAHN, A., *Médecine sciences*, J.-M. ROBERT, IV, 157, 1988.

KOURILSKY, P., *Les artisans de l'hérédité* [The artisans of heredity*], Odile Jacob, 1987.

RIBES, B., *Biology and Ethics*, Insights No. 2, UNESCO, 1978.

ROBIN, J., *Changer d'ère* [A change of eras*], Le Seuil, 1989.

STUPFEL, M., *Environnement et médecine* [Environment and medicine*], Privat, 1986.

TETRY, A., *Jean Rostand, un homme du futur* [Jean Rostand, a man of the future*], La Manufacture, 1988.

WATSON, J.D., *The Double Helix*, Macmillan, 1968.

* These references have not been published in English.